# How Green Became Good

# How Green Became Good

**URBANIZED NATURE AND THE
MAKING OF CITIES AND CITIZENS**

Hillary Angelo

The University of Chicago Press   CHICAGO & LONDON

The University of Chicago Press, Chicago 60637
The University of Chicago Press, Ltd., London
© 2021 by The University of Chicago
All rights reserved. No part of this book may be used or reproduced in any manner whatsoever without written permission, except in the case of brief quotations in critical articles and reviews. For more information, contact the University of Chicago Press, 1427 E. 60th St., Chicago, IL 60637.
Published 2021
Printed in the United States of America

30 29 28 27 26 25 24 23 22 21 20     1 2 3 4 5

ISBN-13: 978-0-226-73899-4 (cloth)
ISBN-13: 978-0-226-73904-5 (paper)
ISBN-13: 978-0-226-73918-2 (e-book)
DOI: https://doi.org/10.7208/chicago/9780226739182.001.0001

Library of Congress Cataloging-in-Publication Data

Names: Angelo, Hillary, author.
Title: How green became good : urbanized nature and the making of cities and citizens / Hillary Angelo.
Description: Chicago : University of Chicago Press, 2021. | Includes bibliographical references and index.
Identifiers: LCCN 2020037021 | ISBN 9780226738994 (cloth) | ISBN 9780226739045 (paperback) | ISBN 9780226739182 (ebook)
Subjects: LCSH: Urban parks. | Public spaces—Environmental aspects. | Nature and civilization.
Classification: LCC SB486.S65 A544 2021 | DDC 333.78/3—dc23
LC record available at https://lccn.loc.gov/2020037021

*For my father*

CONTENTS

Introduction: Urban Greening beyond Cities     1

PART 1    *Green Becomes Good*

1   The Imaginative Turn to the City     29
2   Building an Urban Future through Nature     54

PART 2    *Contested Social Ideals*

3   The Space-Time of Democracy: Parks as a Bourgeois Public Sphere     77
4   Proletarian Counterpublics: Reimagining the Colonies     107

PART 3    *The Social Life of Urbanized Nature*

5   Producing Nature, Projecting Urban Futures     141
6   Experiencing Nature as a Public Good     172

Conclusion: Global Greening Today     201

*Acknowledgments* 219
*References* 223
*Index* 247

INTRODUCTION

# Urban Greening beyond Cities

Since 1994, New York City's Brooklyn Botanic Garden has sponsored an annual "Greenest Block in Brooklyn" contest that "promotes streetscape gardening, tree stewardship, and community development" by encouraging residential and commercial blocks to plant flowers and add greenery to front gardens, window boxes, and tree pits ("Greenest Block in Brooklyn," n.d.). The 2014 winner "scored a perfect 100% on participation" and turned "neighbors" into "friends" (Love 2014). The contest rewards such activity because, as the *New York Times* recently reported, plant material is an index of "civic spirit" and "community cohesion." "It boils down to people more than plants," said the program manager. "It's where you get that tangible sense of love. Which block really felt the most unified? Where is that sense of community really most palpable? That's urban resilience right there" (Levine 2017).

At first glance, there is nothing particularly unusual about these sentiments. In an era of climate change and ecological crisis, urban greening is understood to be a worldwide policy trend, and the ideas motivating the Greenest Block contest are a recognizable global phenomenon. City planners, architects, activists, and municipal governments all use green space to make urban environments more hospitable and sustainable. Scholars study its political, social, and economic effects on urban neighborhoods: perceptions of urban gardening as a "transformative" and counterhegemonic activity (White 2011); how people "use nature as a source of moral value" (Bell 1995, 120); the fact that greening initiatives "richen" and "whiten," their desirability increasing property values and, thus, contributing to gentrification (Gould and Lewis 2016). Green and sustainable cities have become paragons of ideal urbanism, not only in the

West, but also in new "smart" cities in the Middle East and large "megacities" in the Global South. In China, "urbanism of ambition . . . has a green face" (Sze 2015, 14), while Abu Dhabi's Masdar City sees itself as a global prototype of ecocity futurism (Günel 2019, 204). Urban greening—undertaken in the name of ecological sustainability and resilience as well as quality of life—is being carried out in a wide range of places with very different histories, local ecologies, and available aesthetic and ecological repertoires for urban life.

But why should plant material be so widely interpreted as a tangible indicator of "civic spirit" or a "sense of love"? The addition of green space to cities is not intrinsically meaningful. There is nothing inherent to the activity that should cause greening activities to be interpreted as a sign of community spirit or care for a neighborhood and encouraged by contests like the Greenest Block, even and especially in cases in which such interventions might be at odds with ecological goals. In the absence of these associations, not only would it not necessarily occur to people to "improve" urban environments with street trees, flowers, or window boxes; these actions would also not be interpreted by audiences as improvements. And though it is not too difficult to think of other things people do to improve urban environments—like fill them with public art, or optimize their efficiency by rationalizing street plans, or guarantee all their residents access to housing—it takes real effort to imagine a counterfactual reality in which greening acts carry no symbolic significance, one in which planting city streets with trees and flowers would be seen as foolish, or decadent, or even a public health risk.

These practices and associations are the object of analysis of this book: a "green-as-good" logic that is endemic to the planning, design, and study of urban green space; the set of practices celebrated by the Greenest Block contest that are common, not only in large cities in Europe and North America, but in a wide range of environments across the world today. The challenge to imagine green as a neutral value underscores the historicity and contingency of associations of green with good. And, if there is nothing necessary, inevitable, or permanent about this configuration, greening's recurrence as a recognizable, aesthetic and moral practice over time and across place is a thing to be explained. When people want to make urban environments better they often turn to nature. But under what conditions is it possible to use everyday signifiers of nature to improve urban environments? What are the structural, material, and social or cultural conditions for greening to be legible as a form of moral action in the first place? And what are the effects of this recurrent activity on urban politics and urban transformation?

I define *urban greening* as the normative practice of using everyday signifiers of nature to fix problems with urbanism. In urban practice, the catchall term generally includes superficially "green" forms of streetscaping—urban farms, community gardens, street trees, pocket parks—as well as "gray" high-tech ecological interventions such as green building and coastal adaptation. I use this capacious popular policy term deliberately, to highlight the commonalities across this range of present practices and the continuities with those of prior eras. If, in the form of urban sustainability and climate adaptation, greening appears quintessentially contemporary, in another sense urban sustainability is but the most recent iteration of green urbanism. Raymond Williams (2005, 71–75) famously quipped that *nature* is one of the most complex words in the English language, connoting urban parks and apparent wildernesses, plague and famine, providence and destruction, force and resource. In this sense, as long as there have been cities, there have been efforts to improve them with desirable symbolic forms of nature, such as parks and green space, alongside as many efforts to tame, protect them from, and ameliorate the effects of extreme weather, wild animals, and ecological threats of various kinds. The objects, practices, and associations of interest here are those commonly understood to have originated in nineteenth-century urban-industrial environments: the widespread recognition of plants, trees, small animals, and especially green open space as beneficial investments in the public good, understood to be important ways to show care for a city and its people.

Thus defined, urban greening is the very opposite of rare or momentous. As a social phenomenon, its significance is its ubiquity and durability: the common sense of these associations and practices over time and across place; the fact that greening is not simply available as a solution to problems with urbanism but is actually selected quite often by a wide range of actors, from urban boosterists to guerrilla gardeners, in a wide range of urban environments, from large metropolises to smart cities to leafy suburbs. As a consequence, greening is also materially significant. To the extent that this logic is deployed when making decisions about the built environment, greening projects transform cities: shaping living, working, and leisure environments, the organization of public and private space, and, thereby, social organization and civic life.

How might we best understand urban greening as a contemporary global phenomenon? As Julie Sze has put it: "That Shanghai . . . has a green veneer needs serious interdisciplinary analysis that takes space and time seriously" (2015, 14). Global greening begs the question, Why is it that places where urban form does not require it and with such different histories, morphologies, and ecologies are recognizably greening today? How might we explain why, not only in New York in 1850, but also in New York in 2020, as well as in countless shrinking cities and sprawling suburbs across Europe and the North American Rust Belt, and in smart cities of the Middle East, the idea that greener cities are better cities has been naturalized as everyday common sense?

These questions are difficult to answer in part because the study of urban greening has been dominated by well-known cities at notorious moments of urbanism—paradigmatically, Central Park, a large public space, in nineteenth-century New York City, a large industrial metropolis. As a result, the greening impulse has classically been understood as a reaction to the pathologies of large industrial cities—their slums, density, and rampant public health problems (Hall [1988] 1994)—with projects often led by a cosmopolitan middle class engaged in conspicuous consumption (Green 1990) or progressive reform. But accounts of greening derived from such landmark places and moments are misleading. Understanding greening as a reaction to urban pathologies and density can explain greening in New York, the stereotypical concrete jungle. But, in placing explanatory weight on the physical conditions and urban form of a specific time period, such explanations miss the extent to which greening is bound up in social processes playing out across broader spatial and temporal scales and, as a result, offer few resources for understanding how and why this idea and these practices have been able to travel so widely and remained so compelling in such a range of environments today.

This book takes a different approach, developing an explanation for contemporary greening through a historical study of greening projects in Germany's Ruhr Valley, a classic urban-industrial region well-known for its coal mining and steel production, but a place that is failed by traditional explanations of why cities are greened. As a result of its industrial economy, the Ruhr is often invoked as a region without nature—its natural resources extracted, its air and landscape polluted. But, in fact, it has recurrently greened throughout its urban history in the absence of the classic urban form, urban problems, or urban bourgeoisie typically understood to motivate greening. Because the polycentric Ruhr remained low density throughout its industrialization and preserved ample open space—first in the form of farmland, then greenbelts, and most recently recreational greenways and blueways—it never lacked the signifiers of nature—grass, trees, and open space—that are often added to large metropolises. What is interesting about this case is that, even though the Ruhr was and is superficially green *looking*, it could still be "greened"; an empirical lack of open space was not necessary for signifiers of nature to be legible. At the beginning of the twentieth century, industrial barons provided garden cities for workers already living in faux-agricultural environments; postwar planners designed large regional parks to rebuild urban public life; in the past few decades, this deindustrialized region has been made into a giant regional park showcasing industrial nature (*Industrienatur*) and industrial culture (*Industriekultur*). Because, over 150 years, changing groups of protagonists recurrently used signifiers of nature to "improve" the region even in this low-density landscape, the Ruhr's long urban history and unusual morphology make it a place well suited for developing a different explanation—one that must account for greening's origins in an apparently anomalous location and that is, therefore, better able to explain the ubiquity of greening across a wide range of contemporary environments.

From the history of greening in the Ruhr, I argue that urban process, not urban form, turned signifiers of nature into "goods" for society and that urbanization as a social process is the condition of possibility for acting to improve urban environments through green space. Specifically, I find that urban greening is a social practice made possible by a social imaginary I call *urbanized nature*, that was itself an outcome of and that has subsequently become a variable in urbanization. Through historical analysis, I also identify characteristic logics with which this practice unfolds across different moments and projects. Scholarly portraits of greening regimes tend to emphasize their situatedness in a given context: greening appears as an elite, "civilizing" process in the nineteenth century, as large-scale Fordism in the twentieth, or as neoliberal greenwashing and spectacular park development in the twenty-first. This book provides historical accounts of greening projects in these three periods as embedded in these international trends, but it argues that, across this concrete variability, nineteenth-century urban parks and contemporary green cities are—formally—products of the same social imaginary. Centrally, I argue that greening is a mode of remaking cities, socially and spatially. It is also a particularly powerful way of intervening in the urban built environment because, although specific projects are embedded in the political economy of each moment and reflect its biases, in each era they are constructed as universally beneficial investments in the public good by both greening protagonists and their target audiences. The purpose of this book is to bring this view of greening—as a "grammar" of moral action (Mische 2014), as a favorite fix for problems with urbanism—into focus by treating the built environment as an archive of these decisions.

One note before we begin: The words *green*, *nature*, *good*, and *improvement* appear frequently throughout this book. Because the use of scare quotes has been minimized, I ask readers to imagine them when you see such words used. These are descriptive terms, used to refer to everyday understandings of the material referents and moral associations that are the objects of analysis of this book, rather than analytical terms used to make ontological or normative claims about ecology, society, or urban reform.

## EXISTING ARGUMENTS AND THEIR PROBLEMS

Urban and environmental historians have generally explained greening as a reaction to the morphology and social problems of industrial cities by the urban bourgeoisie, a philanthropic, nature-appreciating class that embarked on a variety of nature-oriented social reform projects such as investments in public parks and weekend excursions to the countryside (e.g., Bender 1982; Green 1990; Nash 2014; and Schmitt 1990). Though aspects of these sensibilities can be traced to antiquity (Bell 2018; Glacken 1967), contemporary views of nature as a good for society are commonly understood to be products of modern urban life; as the historian Roderick Nash has put it, "nature appreciation" is a "full stomach" phenomenon: "Society must become technological, urban, and crowded before a need for wild nature makes economic and intellectual sense" (Nash 2014, 44, 344).

This explanation is inadequate for several reasons. First of all, it fails empirically. Suggesting that the morphology of large industrial cities is the catalyst for greening cannot account for the recurrence of the practice across different places and times or its use by a multiplicity of actors with varying relationships to power: why not just New York but also cities like Abu Dhabi are recognizably "greening" today. But, more importantly, it is an explanation that is produced by and that reproduces two long-standing polarities in urban studies—between the study of the city and of nature, and between culture and political economy.

Let us begin with the city and nature. The idea that greening is a reaction against the city suggests that greening projects are spatializations of city/nature (or society/nature or culture/nature) dichotomies—that cities are human-made concrete jungles, that nature is an other that can be brought in to improve the social world, that urban green spaces are oases from urban life and their social relationships. This formulation reproduces classic representations of nature—and, by extension, the countryside—as foundationally opposed to modern urban life that have long been at the core of modern European urban and social scholarship as well as of popular imaginaries (Čapek 2010; Park and Burgess [1925] 1984; Wirth 1938). This view was largely a product of industrial urbanization in the nineteenth century (Wachsmuth 2012; Williams 1973). Growing cities appeared to devour the countryside and its resources; industrialization tore people from the land, moved them to cities, and exposed them to larger and more complex social worlds. As the first generation of European and American sociologists examined the social consequences of these transitions, new characteristics of urban society were understood as analogues of these spatial relationships. The country was the home of nature, while the city was the home of society. Thus, urban *Gesellschaft* (society) implied alienation from nature—or at least gave no mention of this relationship—while rural/agrarian *Gemeinschaft* (community) was at least implicitly linked to agricultural economies, land ownership or tenure, and physical location in the countryside (Simmel [1902] 1964; Tönnies [1887] 2011). These binaries were subsequently transposed onto the organization of disciplinary fields—urban studies and rural studies, urban studies and development studies, sociology and ecology (see Angelo 2017; Dunlap and Catton 1994; Robinson 2002; Smith 2011)—with the result that, until recently, sociology could ignore nature, just as environmental studies ignored the city. A classic example of this opposition is the Chicago school's erasure of Lake Michigan in Burgess's (1925) concentric zone model of urban growth.

A second persistent polarization is between what we might call *cultural* and *materialist* approaches to the study of nature in the city. It is from the materialist perspective that greening can appear as a reactionary mode. Understandings of greening as a response to urban problems suggest both a certain temporal order—that urban problems come first and green solutions second— and an implicit or explicit critique of such efforts as ideological in the strict, Marxist sense: as a misrecognition of the problems (and possible solutions) of industrial capitalism and as a reflection of dominant class interests (Purvis and Hunt 1993; Williams 1977, 108). Such a view risks placing greening in the realm of the superstructural and/or epiphenomenal. Though, as David Harvey has noted (not coincidentally, also in the context of an argument about ideas of nature), "even Marx" was "willing to countenance ways ideas could become a 'material force' for historical change when embedded in social practices" (Harvey 1993, 31), and in spite of the long history of efforts to bridge structural variants of Marxism with more cultural approaches outside specifically urban-environmental work (including those by scholars such as Raymond Williams, Henri Lefebvre, and Cornelius Castoriadis on which this book draws), this cultural and social work has historically been overlooked and/or undertheorized in critical geography (Smith 1998, 277). Environmental sociologists, meanwhile, have followed anthropologists in taking a more Durkheimian approach, emphasizing nature as a mirror for culture (Geertz 1972) and, often, as a site of social or moral conflict (Bargheer 2018; Farrell 2015) or a place where cultural, ethnic, or class differences are played out (Bell 1995; Douglas [1966] 2003; Fine 2009). While well attuned to the environments and practices that produce particular nature sensibilities, sociological approaches have not always situated these in relationship to broader transformations of urban space or urban political economy.

As a social phenomenon, then, urban greening has been inadequately understood as a result of these antinomies, by scholarship on the topic that is often organized to reflect them. This book is part of a large and growing body of work that is situated at their intersection and that has been bringing a new understanding of the material and social relationship between cities and the natural world to urban scholarship and practice in the past several decades. Urban-environmental thinkers have reversed the modernist narrative of cities as places without nature, addressed entanglements of green materialities and imaginaries, and redescribed urbanization as a process of large-scale "socionatural" transformation (Cronon 1992; Gandy 2002; Heynen, Kaika, and Swyngedouw 2006; Kaika 2005; Smith [1984] 2010). Recent studies of urban infrastructure, ethnographic and comparative accounts of human-animal relationships, and environmental histories of capitalism have all challenged conventional understandings of apparently social and natural environments (Ernstson and Swyngedouw 2018; Jerolmack 2013; Moore 2015; Tsing 2015).

Urban greening has not yet received this treatment. Classic explanations for greening as a reaction against the city do not reflect these current understandings, while contemporary reappraisals have so far devoted less attention to the causes and consequences of the quotidian practice of bringing signs of nature into cities. In addition, even as scholars and practitioners increasingly argue against inherited city/nature dichotomies, there has been little sustained engagement with questions of greening's role in remaking cities—what it means, analytically and methodologically, to center urban green spaces as key domains through which to communicate new ideals of cities and citizenship and as pedagogical spaces where urban norms and habits are learned and reproduced. Particularly now—as greener cities are prescribed as solutions to climate change and in the context of the widespread commodification of greenness in contemporary capitalism—a view of nature's imaginative power is central to understanding its role in urban political economy.

The goal of this book is to provide an explanation for urban greening that overcomes these polarities and reflects recent theoretical advances. The objective is to move beyond the two persistent binaries—between city/nature and culture/political economy—reflected in commonsense understandings of greening as an (ideological) reaction to the (industrial) city by developing an explanation for greening that foregrounds the relationship of urbanization as a social process to greening as a social practice, that highlights the centrality of green space to the organization of modern urban life, and that foregrounds the causal role of cultural or imaginative work in relationship to structural and material factors, all in order better to understand the dynamics and consequences of urban greening today.

## *RUHRBANITÄT*

I develop this alternative explanation by studying a place failed by existing ones. Germany's Ruhr Valley is one of Western Europe's largest urban agglomerations, its seventeen hundred square miles containing eleven cities, four administrative districts, and over seven million people in North Rhine–Westphalia (NRW), on the western edge of Germany. The Ruhr was early to industrialize, its coal mines and steel mills operated through both world wars and two postwar occupations, and today it possesses a strong working-class identity rooted in its industrial past.

The Ruhr is also a region full of paradoxes. Historically known for its coal mining and steel production, today it is most famous for the Internationale Bauausstellung (IBA) Emscher Park, a ten-year, regionwide reinvention of its brownfields as parkland, industrial relics as museums, and rail lines and sewage canals as greenways and blueways that followed the collapse of its industrial economy, culminating when "Essen for the Ruhr" was named a European Capital of Culture in 2010. The Ruhr is also a case of urban greening in the absence of a city. Though it has always been low density and never lacked open space, its urban history has been marked by recurrent efforts to turn a working-class region into a legibly middle-class city through greening projects. Planners used garden cities and greenbelts to create its first urban publics and a legible urban form following its rapid industrialization early in the twentieth century, designed large regional parks to rebuild urban public life in postwar West Germany, and, in the late twentieth century, turned industrial landscapes into green leisure spaces to enhance regional competitiveness following the collapse of the industrial economy.

The Ruhr is a strategic site in which to examine urbanization as a process of social transformation and, specifically, the relationship of this process to greening practices because it is an exemplar of noncity urbanism. The polycentric region lacks—and has long lacked—a traditional city. It grew out of a chain of medieval market towns and, for 150 years, has been seen as a problem urban environment, lacking both a classic city form and a cosmopolitan citizenry: it is spatially diffuse and culturally provincial. It is not one large city but consists (depending on how its boundaries are drawn) of about seventeen hundred square miles containing (depending on how you count) ten to twenty to even forty-five cities, four administrative districts, and five to eight million people. Its social composition is also unusual in that it has historically lacked an urban bourgeoisie. Like Pittsburgh or Detroit in the United States, the majority of its population was, until recently, employed in mining and manufacturing, and it has been best known and most often represented in Germany as polluted and working-class. In the absence of familiar social and spatial markers of urbanism, it is also an experientially confusing landscape. The German urban sociologist Walter Siebel once observed: "If one asks people in Europe what they immediately associate with the term urbanity, there is little by way of hesitation. Urbanity means crowded streets, 24-hour shopping and the traditional character of the European city." But in the Ruhr: "Everything that fits in with the image of a European city is missing—a central core, urban/rural contrast, and a mix [of functions]. This is urbanized countryside without a real city" (1999, 123).

Though local practitioners and elected officials have seen these characteristics sometimes as problems and sometimes as opportunities, they have made the Ruhr a fertile ground for urban scholarship. Friedrich Engels wrote on Ruhr miners' strikes in the nineteenth century, and for Max Weber the towns of Essen and Bochum were models of the producer city. In the twentieth century, German social historians of everyday life—sharing methods and politics with Raymond Williams, Henri Lefebvre, and above all E. P. Thompson—carried out much of their research in the region (Eley 1989; Lüdtke 1995). Attempts to make sense of the region's polycentric urban form—which has long resisted efforts to characterize urbanism in terms of discrete settlements—have almost as long a history. In 1912, the prescient local planner Robert Schmidt argued that, when the Ruhr was compared to traditional industrial cities, its low-density polycentrism made it a model of "sound, healthy, and beautiful" urban structure (Schmidt 1912a, 42; see also Von Petz 1990, 9). When Peter Hall's *The World Cities* was published in 1966, the book listed seven: London, Paris, Randstad (Holland), Moscow, New York, Tokyo, and the Rhine-Ruhr. More recently, the urban theorist Thomas Sieverts developed his concept of a *Zwischenstadt* (which translates literally as "in-between-city") largely in reference to the Ruhr (Sieverts [1997] 2003). In 2013, the Technische Universität Dortmund even organized a public lecture series and planning seminar devoted to the perennial question of the Ruhr's unique urban form and culture. The series was titled "Ruhrbanität," a pun on the German word

*Urbanität* that we might translate as *Ruhrbanity* (see Reicher, Kunzmann, et al. 2011).

As must already be clear, in this analysis the Ruhr figures as an *urban*, not a *German*, case. Work on representations of nature as spatializations of social and political power has tended to take the nation as its context (Carroll 1996, 2002; Fourcade 2011; Mukerji 1997; Scott 1998). Chandra Mukerji, for instance, has described the gardens of Versailles as "a form of material practice, a way of acting on the land that helped to make it seem like France" (1997, 9). And, of course, other imaginaries of nature—especially antiurban and antimodern ones—featured prominently in state projects and played out in the landscape in Germany in particular.* In the projects examined here, greening was a way of acting on the land to help make the Ruhr seem like a *city*. Though the projects reflected specific national concerns, such as rebuilding democracy in postwar West Germany, greening protagonists self-consciously situated themselves in an international comparative field of other cities and other urban actors. And, though the projects met utilitarian needs for residential and recreational space, they were also understood as demonstrations that were part of deliberate attempts to remake the Ruhr to reflect new urban ideals. The ideals to which protagonists aspired in each moment were also self-consciously urban as opposed to national. They were modeled after international trends and reflected both recognizable international crises of urbanism and paradigms for the design and ascribed social functions of urban green space.

If the Ruhr is analytically useful because it is an unusual historical case, the explanatory challenges it poses also make it instructive today as a typical one. The great diversity of twenty-first-century urban forms and

---

* The relationship between nature and nationalism in Germany is particularly fraught, given the Nazis' use of nature tropes and the connections drawn between race and landscape. But, from romanticism to the rise of contemporary environmental movements, there is a long tradition in Germany of ideas about nation, nature, and urbanization (e.g., Cioc 2002; Lekan and Zeller 2005) that intersects with the transformation of industry and the environment in the Ruhr (Brüggemeier and Rommelspacher 1992). Figures such as the composer Richard Wagner and the painter Caspar David Friedrich typify this tradition—Friedrich's wild, windswept landscapes show humans confronting the full power and pathos of the wilderness. These sentiments carried over into the nineteenth century's *Naturschutz, Heimatschutz, Landschaftspflege*, and *Lebensreform* movements, which, in the name of both spiritual health and national heritage, eventually promoted hiking and outdoor activities as well as the preservation of folk culture and rural landscapes. The international urban imaginary being studied in the Ruhr emerged even in the context of this strong national culture of romantic, antimodern, rural or wild nature and coexisted, historically, with those practices and ideas.

ecologies—Midwestern suburbs, tropical megacities, desert megalopolises—and the presence of appeals to greenness across them beg an account of the travel, legibility, and polysemy of these signifiers. The Ruhr's recurrent greening in the context of ample green space challenges traditional explanations rooted in particular morphologies and requires a new and different one. If not the experiential contrast of apparently green nature against an apparently gray city, what *are* the conditions of possibility for greening? And, even if the practice arose in dense, centralized industrial cities, how did greening travel, and why has it remained so ubiquitous across changing urban conditions—across three time periods in an unusual location and, by extension, across a wide range of urban environments with very different histories today?

## A HISTORICAL SOCIOLOGY OF URBAN GREENING

If green-as-good sentiments are not inevitable but learned and reproduced, it should be possible to trace their travel and take-up in a given location. In New York, for instance, the online amateur historians the Brownstone Detectives have traced the origins of "block beautification" to 1902, when "a number of private Brooklyn citizens . . . initiated an organized effort" to green their neighborhood. Newspapers report that, after initial skepticism, the idea "spread," catalyzing the organization of block associations, a "record" tree-planting initiative (sixty-three were planted, one for each residence), and the eventual "co-optation" of the idea by realtors to market their properties and attract tenants ("The Greenest Block in Brooklyn [1902]" 2017).

Because the aim here is not only to document urbanized nature's arrival in the Ruhr but also to arrive at a more general explanation of this social imaginary's origins and effects, this study has been designed as a longitudinal comparison (pace Walton 1992) of one place over time. This method, designed to highlight similarities rather than differences, makes it possible to identify recurrent patterns in greening's social logics and consequences. It is also a historical sociology as defined by the British cultural materialist Raymond Williams, which he charged with the explanatory task of giving a "theoretical account of the *conditions of a practice*," exploring the "relations between these material means and social forms and the specific (artistic) forms which are a manifest cultural production," and examining a particular art form or cultural practice as *cultural* in the sense of a "signifying system" converging with a "whole way of life" (Williams 1995, 145, 88, 13).

Williams undertook such a project in his classic *The Country and the City* (1973), which analyzed three centuries of British literature to explain nostalgic depictions of the idealized countryside in contrast to the city as a place of suffering, inequality, and discontent. As it was for Williams, in this study the historical perspective is a way of better understanding present practices. Analyzing a century and a half of greening projects makes it possible to give a "theoretical account" of the conditions of possibility for greening, a historical account of the relationship between these "material means and social forms" and green cultural projects of each era, and a sociological account of greening's logics that explains how this particular "signifying system" converges with urban life and politics.

To establish under what conditions it is possible to select nature as a solution to problems with urbanism in the first place, I trace the origins of greening in the Ruhr in the wake of rapid industrialization and urbanization in the nineteenth century. To develop a sociological account of greening's uses and effects—or, in Williams' terms, to explain how this signifying system converges with a "whole way of life"—I also compare greening projects across three moments of major global urban restructuring: industrialization at the end of the nineteenth century, deindustrialization in the middle of the twentieth, and neoliberalization at the beginning of the twenty-first (Aglietta 2000; Brenner 2004; Jessop 2002). Concretely, in each period I profile two major projects that were both locally significant and embedded in and reflective of internationally recognizable greening paradigms of each moment: the nineteenth-century garden city in the first decade of the twentieth century, the postwar city beautiful (and New Left reactions against it) in the 1970s, and the entrepreneurial green city in the early twenty-first century. The book documents the changing ideals of urban society and the ways in which greening projects were used to remake the Ruhr's cities and citizens in each moment in the service of a sociological project: I use this comparison to identify common logics with which greening projects unfold aross the differences of these three periods.

The significance of these moments of urban restructuring is twofold. First, as periodization. These were moments in which greening protagonists were self-consciously focused outward: they were responding to international moments of urban crisis with local greening projects that reflected the trends and structural constraints of each era. Second, scholars understand crisis moments to be moments of social creativity—times when societal norms are up for debate and new repertoires of social action emerge (Sewell 1996). This study of greening projects in such moments highlights how their proponents conceived new designs for public and private green space as rearrangements of living and leisure, and how their interventions were oriented toward realizing new ideas of the good life. In so doing, it also positions these greening projects as sites of well-intentioned action: protagonists were not just deriving instrumental solutions but, in each moment, driven by normative visions of ideal cities and citizens.

Ultimately, this book contains two periodizations. First, there is the periodization of *urbanized nature*: the horizons of possibility within which it is possible to green in the contemporary, recognizable sense, a process that has, I suggest, a longer history and a wider spatial diffusion than contemporary commonsense understandings imply. Second, there is a periodization of *urban natures*: the portraits of the specific physical, aesthetic, social, and political forms that urban greening takes at each moment as it is used to construct different ideals of cities and citizens, and across which I identify formal similiarities.

Data for this project were collected during two years of fieldwork, from 2011 to 2013, including one year living in the Ruhr city of Dortmund. I carried out twenty-two formal and semistructured interviews with municipal and regional government officials and urban planners and engaged in approximately two hundred hours of participant observation at events, art exhibitions, community meetings, lectures, and conferences and in museums, parks, and public spaces. I visited several local archives, including that of the Regionalverband Ruhr (RVR, the regional planning association) and the personal archive of Roland Günter, one of the leaders of the movement to save the colonies profiled in the postwar period. Evidence presented here about the rationale for the projects, especially regarding how they were seen at the time to align with international trends and expected to catalyze change locally, is drawn almost exclusively from these sources. In addition to original archival materials, I also relied on the many excellent social histories of the Ruhr region—especially regarding its nineteenth-century colony housing and employer/employee relationships—rereading them through the framework developed here. Finally, but significantly, scholarly writing that directly informed the projects during the time of their construction is used as a historical source. The closer relationship between urban theory and urban practice is unique to the German case—at least in comparison to the United States—but works by Jürgen Habermas, Oskar Negt, and Alexander Kluge in the postwar period and by urban theorists such as Thomas Sieverts in the IBA case are cited as influences on the projects only when practitioners made it clear that they were.

## AN ALTERNATIVE FRAMEWORK: URBANIZED NATURE

This approach produces an account of urban greening not as a reaction to morphological attributes of particular cities but as a social practice that is an outcome of, and a variable in, urbanization as a global process of sociospatial transformation. But *urban*, of course, is a word nearly as complex as *nature*. What does it mean to attribute these ideas and practices to urbanization?

## The Urbanization of Ideas about Nature

I call the imaginary that motivates greening practices *urbanized nature* to distinguish it from *urban nature*, a term commonly used to refer to "nature in cities" or "first natures" transformed by urbanization (Angelo 2017; Angelo and Wachsmuth 2015). Urbanized nature is meant to signal that what makes these sentiments and practices urban is not that they take place in cities (though they may) but that they are outcomes of urban transformations.

While the extraction of natural resources, the degradation of apparent wilderness environments, and the ways in which nature acts back on society—for example, through flood, fire, and pestilence—all have significant relationships to urban processes and merit books of their own, these forms of nature do not appear in this story. Instead, the object of analysis is an imagined extrasocial nature that is recurrently brought in to improve the social world. I use *urbanized nature* to refer to *ideas about nature* that have been transformed by urbanization and focus on the set of everyday symbols of domestic "second natures" that are legible as goods for society: parks, trees, small animals, gardens, and green space. In this book, the word *city* is used to describe a "category of practice"—the normative ideal of society to which greening protagonists aspired—and *urbanization* a "category of analysis" (Brubaker and Cooper 2000): the bundle of transformations in productive, spatial, and social/cultural relationships that set the stage for these new ideas of nature to emerge, the paths along which those ideas traveled, and the field that they aimed to influence.

This orientation is intended to highlight the significance of these representations and social uses of nature in affecting the actual coconstitution of society and nature in a given era. It does so by centering urbanization as a large-scale process of global historical transformation, the significance of which is often lost in binary accounts focused on cities as discrete places and/or a nature outside them. This book is thus part of recent calls in critical urban studies for a reorientation from place to process—from a focus on cities as particular settlement types to one on urbanization processes (Angelo and Wachsmuth 2015; Brenner 2013; Brenner and Schmid 2015). The core argument is built up from Lefebvre's ([1970] 2003) diagnosis of a coming "complete urbanization" of society and his prediction that it would soon no longer make sense to speak of cities as such, that we should, rather, speak of a differentiated "planetary" urban fabric. The key tenets of such a perspective as a contemporary research program for urban studies are (1) a shift from morphological form and attributes to historical processes of sociospatial transformation (from cities to urbanization) as the proper objects of urban analysis and (2) an emphasis on multiscalar and multisited explanations of urban phenomena (Angelo and Goh 2020). To advocate for this view is not to suggest that all places have become cities or that all cities are now just like each other but to argue that these processes have affected the entire planet—differently in different places—just as have other processes of global historical transformation, such as colonialism and capitalism, and to suggest that urbanists take a wider view of particular places, cities, and problems as bound up in these global transformations and as part of their explanatory remit. An approach guided by planetary urbanization would, for instance, insist on a view of extractive landscapes like Alberta's polluted tar sands as intrinsically linked to sites of urban concentration, such as Vancouver, that rely on these natural resources (Brenner 2014).

Such accounts also open up the possibility for theorizing urban modes of perception and ways of life, traditionally understood to be products of cities, as outcomes of urbanization processes as well (Angelo 2017). To date, arguments for the study of urbanization as a multiscalar, socionatural process have emphasized its spatial dimensions, especially the physical transformation of cities as linked to broader extractive landscapes and global political economies (e.g., Arboleda 2016; and Brenner and Katsikis 2014). Words suggesting the social and experiential aspects of these transformations—such as *epistemologies* and *imagination, phenomena* and *condition*—frequently appear in landmark texts, and it is presumed that, just as urbanization processes move and transform goods, people, and raw materials around the globe, cultural forms, practices, and ideas circulate as well. But what does it mean to specify urbanization as process that works on ideas about nature? What are those cultural forms and practices, and how is it that they travel?

This book is about the social, cultural, and experiential dimensions of urbanization, not in terms of the phenomenal experience of the city as a particular kind of morphological environment, or of the experience of "complete urbanization" outside recognizable city/country distinctions (Schmid 2014), nor of the concrete diversity of urban experience at the level of everyday life, but as transformations that are outcomes of a large-scale process of historical transformation. It is about the urbanization of social consciousness understood in terms of

- The role of global urban transformations—in social, spatial, and economic relationships—in producing categories of practice, especially those usually

taken to be products of city experiences and associations, such as understandings of and moral attitudes toward nature.
- The behavioral and cognitive effects of being materially and socially/culturally embedded in an international group of cities and urban actors that becomes a reference group and competitive field for local norms, policies, and decision making.
- The use and social legibility of the categories and associations produced through these entanglements in a variety of urban environments, not just in large cities or areas of agglomeration.

This perspective yields the following argument. It was not a specific set of physical conditions that produced greening (dense, centralized cities per se). Instead, a broader set of underlying transformations, especially the shift from agricultural to wage labor and from subsistence living to a market economy, changed forms of engagement with these materials and spaces from labor to leisure: from involvement with plants, animals, and green space as "direct" material goods required for subsistence purposes to "indirect" or moral or affective goods. Once available—initially in Europe and North America—this view of nature made it possible to use greening projects as solutions to problems with urbanism in large industrial cities, like New York, but also beyond them, in places like the Ruhr in the twentieth century or Asia or the Middle East today. In addition, the Ruhr's turn to greening suggests that wider-spread social transformations may have been necessary but were not sufficient conditions for greening to take hold in any specific place. Greening practices (and the city as an ideal to which to aspire) are not automatic outcomes of urbanization. Instead, at least in some cases, a more immediate catalyst was protagonists in a given location coming to be—and being self-conscious of being—economically, politically, and socially embedded in an international network of other cities and urban actors through which this view of nature spread.

To put this in more abstract terms, we might turn to Neil Brenner's distinction between the urban as *nominal essence* and as *constitutive essence*. *Nominal essence* refers to the social and spatial forms that urban "phenomena, conditions, or landscapes" take (e.g., cities themselves), and *constitutive essence* to "the various processes (e.g., capital investment, state regulation, collective consumption, social struggle, etc.) through which the urban is produced" (Brenner 2013, 96, 98). My argument is that this perception of nature was a product of urbanization as constitutive essence: the changing forms of investment, production, consumption, and social relationships of industrial urbanism constituted this new moral view of nature just as they constituted cities.

Once available, urbanized nature has since affected the urban form as nominal essence: cities and other urban environments are constructed and transformed in part through this idea of nature as it is drawn on when decisions are being made about the built environment.

## Urbanized Nature as a Social Imaginary

Of course, I am not the first to link urbanization to transformed ideas of nature or to describe the global diffusion of urban processes and ideas. In 1938, Louis Wirth observed: "The degree to which the contemporary world may be said to be 'urban' is not fully or accurately measured by the total population living in cities. . . . The city is . . . the initiating and controlling center of economic, political, and cultural life that has drawn the most remote parts of the world into its orbit and woven diverse areas, peoples, and activities into a cosmos" (Wirth 1938, 2). The sociologist Michael M. Bell has more recently observed that historically "urban" and "rural" ideas eventually "have an independence of their own," transcending the places and times in which they arose (Bell 2018, 12). But how exactly urban ideas and moral frameworks travel and transform has not been elaborated in recent theories of urban nature, theoretically or empirically. Having equipped ourselves with a framework for thinking about urbanization as a process, we require a second tool to understand the imaginaries of city/nature relationships that urbanization produces, and to move this extant framework into sociological terrain—a concept for a durable, transportable idea; the concept of the social imaginary.

Social imaginaries are cultural understandings of the moral order of a social world that are historically specific and intersubjectively shared within their domain (Anderson 2006; Castoriadis 1997; Taylor 2004).* They are "carried in images, stories, and legends," "shared by large groups of people," and, importantly, "make the reproduction of common practices" possible (Taylor 2004, 23; see also Calhoun, Gaonkar, Lee, Taylor, and Warner 2015). In other words, social imaginaries explain the eternal return of certain practices: why, over time and across place, the same "idiom[s] of expression" (Gaonkar 2002, 10) are invoked to communicate the same families of ideas. Benedict Anderson's classic *Imagined Communities* is perhaps the most famous elaboration of the concept,

---

* Though, in sociology, the term *social imaginary* is often invoked loosely to refer to the interpretive frame that people use to make sense of the world (Jessop 2010; Somers 2008) or envision its future (Baiocchi, Bennett, Cordner, Klein, and Savell 2013; Frye 2012; Mische 2009; Tavory and Eliasoph 2013), I use it in the specific sense intended by its originators.

through the case of nationalism. He showed how, as a collective understanding of the nation as an imagined, finite, and sovereign community arose and "spread" across the modern world, it produced feelings of belonging, drove acts of exclusion, and even inspired people to die for it (2006, 6–7). Beyond nationalism, social forms including states, democracies, cosmopolitanism, the public sphere, civil society, and the market have been shown to be shaped by shared understandings about the rules governing specific worlds, used to evaluate the existing environment, and guide interventions in it (Calhoun 2008; Taylor 2004; Vertovec 2012).

In addition to emphasizing the connection between beliefs and practices, the concept of the social imaginary also highlights the structural and material as well as the cultural and social conditions that allow ideas to travel and take hold in new places. Manu Goswami, in particular, has extended Anderson's account of "modular" nationalism, scoping out from descriptive accounts of the particularities of specific nationalisms or simple models of linear diffusion to examine the thinkability of nationalism as a "social form" in order to address questions of the extent to which given social forms are contagious or how similar idioms of expression are taken up in very different contexts. She argues that material and cultural conditions of global capitalist and colonial restructuring made it possible to think the world in terms of nations, and that the subsequent new "universalizing" and "particularizing" trends produced allowed the concept to travel to very different places, with very different histories, resulting in specific nationalist movements of very different kinds (Goswami 2002, 786–87).

The argument here is an analogous one: that greening functions in much the same way as the performance of nation. As mentioned in the example of the Greenest Block contest, greening is not possible at all times everywhere. In the absence of associations of green with good, adding trees to urban streets bears no obvious moral connotations. But, once these associations are present and widely shared, it becomes possible to "improve" cities through the addition of everyday signifiers of nature (trees, green space) and be confident that these actions will be interpreted as improvements.

Specifically, the argument is that urbanized nature—an outcome of urbanization—is the social imaginary that makes greening practices possible and that accounts for their recurrence. Like the idea of nation, urbanized nature is not a specific idea of nature but a shared understanding of a given set of material representations of nature as specific kinds of goods for society. It is this understanding of nature's benefits and the practices that follow that is transferable and that guides action across greening's various concrete instantiations. Once nature became a recognizable, common idiom of the good, protagonists could green with a naturalized belief in nature as a public good and have their actions be received as such—in large, dense industrial cities but also beyond them—and see these sentiments materially affirmed and reproduced in the built environment. This is not to say that the greener-equals-better formula was or is uniformly held to by all people everywhere; rather, as a result of industrial urbanization in Europe and North America, this perception has become *dominant*: reinforced by institutional and spatial structures and reproduced in public discourse and popular culture (Eliasoph 1998; Gramsci 1996). Urbanized nature can take hold not only in cities but also in places where—thanks to participation in a shared material and imaginative field—it is possible to think the world (or, aspirationally, a given place) in terms of cities.

This account of the relationship between social imaginaries and the social practices they enable also aids in the project to reposition greening as a cause as well as an effect of urban change—a goal that is consistent with the objectives of earlier scholarship using the concept of the social imaginary. Cornelius Castoriadis, who was more focused on questions of social transformation than reproduction, and was especially interested in the origin of original and creative ideas, expressed a similar sentiment in relationship to arguments about structural determinism. Castoriadis essentially considered the social imaginary to be the variance left unexplained by Marxism. He began *The Imaginary Institution of Society* by arguing that it is not just the development of productive forces that "determines" individuals' behavior, and his project to explain the sometimes novel, unpredictable "self-creation of society" was motivated by a political and intellectual commitment to theorizing creative work and outcomes that could not simply be traced to structural origins (Castoriadis 1997, 5, 24–25; see also Gaonkar 2002). Explaining greening as a practice produced by a social imaginary continues to walk this line by showing—in the context of a materialist analysis—how common understandings become conditions for action and have significant material and social effects.

## GREENING'S LOGICS

The empirical chapters of this book document a series of greening projects in the Ruhr over 150 years. This is not a design history, which describes what these spaces physically looked or felt like, but a social history, which documents what they were *imagined* to be like, how they were perceived and described, and especially the social functions that their creators, advocates, and users understood them to serve. Across the differences of each project's actors, agendas, time, and location, through comparison I identify several common ways in which greening is understood in the popular imagination and patterns it produces in public politics. These are as follows.

1. *Signifiers of nature are consistently constructed as indirect, universal, and aspirational goods.* Urbanized nature is marked by three related characteristics. First, rather than a direct good required for subsistence purposes, it is understood as an indirect or moral good in what sociologists of morality describe as a "formal, not substantive," sense (Hitlin and Vaisey 2013, 55; see also Tavory 2011): not in terms of right or wrong, or with reference to any particular goods nature might provide or normative ideals to which green cities might aspire, but as "encompass[ing] any way that individuals or social groups understand which behaviors are better than others, which goals are the most worthy, and what people should believe, feel, and do," the opposite being "not immoral but nonmoral or morally irrelevant" (Hitlin and Vaisey 2013, 55).* In short, urbanized nature is valued, not primarily for the nutrition, shade, or windbreak it provides, but as a container for moral sentiments and as a vehicle for morally charged action.

Second, urbanized nature is understood to be *universal*: beneficial to all and to all in the same way. This is part of the classic ideological construction of the term *nature* beyond specifically urban configurations. Modern society/nature binaries place nature outside the social by definition, whether in the form of sublime "untouched" wilderness or dangerous "wild" nature (Cronon 1995). It is from this perceived separation that nature derives much of what Lorraine Daston and Fernando Vidal (2004) call its "moral authority." If nature stands outside society, it can be sought as respite or escape, but it can also be brought in to improve it.

Third, urbanized nature is *aspirational*. I use this word to signal both the future-oriented and the normative aspects of greening practices. In contrast to traditional accounts of greening as a reaction to problems with urbanism, I find that nature is deployed proactively, with ideal futures in mind. In addition, greening acts are not simply utilitarian solutions to the various problems protagonists confront but deliberate efforts to improve cities and people's lives

---

* As such, this book both extends the growing body of work on the symbolic power of nature and dovetails with the sociology of morality, norms, and values (see Abend 2014; Baiocchi, Bennett, Cordner, Klein, and Savell 2013; Boltanski and Thévenot 2006; Glaeser 2011; and Strand 2015).

by communicating new civic ideals and constructing new civic realities. Much of this imagining precedes physical interventions.

2. *Urbanized nature is an imaginary of form, not content.* This social imaginary is also formal, not substantive: it makes greening into a way of expressing social desires, but its content and politics are highly variable. It is in this sense that greening is better understood as a historically specific idiom or grammar of moral action rather than a specific viewpoint. Urbanized nature is a social imaginary in the definitional sense (shared; agentic; making social action possible), but rather than a specific imaginary *of* society or social institutions—such as markets, states, or nations—it makes a set of material referents into a mode of collective expression, a way of expressing desires, a social form through which to communicate normative visions of society in general.

As a result, greening is a practice that is available to a wide a range of actors and political projects even in the same place and time, and, being available for all kinds of visions, it is also one that has no predetermined relationship to power and hegemony. Put differently, urbanized nature is not an ideology in the strict sense—reflecting a particular class interest and shoring up existing social arrangements—but a vehicle for social visions with various relationships to power. For instance, green spaces are often used for playing out radical alternatives to current arrangements, such as community gardens reclaiming urban space as use value or Occupy Wall Street's use of Zuccotti Park for social critique. While this book primarily documents greening in its top-down, large-scale, and hegemonic moments—and thus represents this diversity only partially—there is nothing inherent to greening practices or this social imaginary that suggests it must reaffirm the status quo.

3. *Greening is a mode of remaking cities spatially and socially.* If, as this book contends, greening is a mode of remaking cities rather than an escape from urban life, signifiers of nature are, more specifically, used to recurrently reconstruct the city as a social world by spatializing changing ideals of publicness in each era. Forced to reformulate a new urban vision for the Ruhr in crisis moments, protagonists tended to do so along two lines—in terms of its ideal spatial form (cities) and its ideal social content (citizens). Those interventions did not produce simply decorative, quality-of-life improvements; they actually determined who would meet, under what conditions, and for what kinds of discussions. Thus, 150 years of greening has transformed the Ruhr's public life by designating who encounters each other where, what activities are made public or private, for individual or collective use, and what kinds of spaces are created for politics.

To call this remaking *creative* and *aspirational* does not mean that it is totally open-ended. Like other forms of historical action—political activism, scholarly thought, etc.—greening is always embedded in social relations, and its concrete instantiations always reflect the material and structural constraints of a given moment. Specifically, I find that greening transforms cities in ways that reflect the "patterning of social life by modes of regulation" in a given conjuncture (Steinmetz 2005, 110) and that in each era greening projects reflect the epistemological conventions, forms of ethics and rationality, and especially the international ideals of urbanism and citizenship characteristic of the political economy of the period. Greening in the Ruhr, for instance, first set out public/private divisions during industrialization at the beginning of the twentieth century, structured a conflict between bourgeois and proletarian forms of public space in the postwar, Fordist era, and created a privatized, income-generating, consumption-oriented public sphere under neoliberalism in the past few decades.

4. *Greening projects are normative projects carried out and received as public goods.* Greening is marked by a paradox, which is that, while greening projects are technologies of control that instantiate very narrow, historically and class-specific ideas about what constitutes good cities and citizens, they are nevertheless carried out and widely received as universally beneficial investments in the public good by both greening protagonists and target audiences. Objectively, the greening projects profiled in the Ruhr are social and therefore partial projects that have repeatedly bolstered dominant bourgeois ideals of middle-class urban life. And, of course, any greening project—whether top-down or bottom-up—will be a space inscribed with the social norms and values of its creators rather than one impartially providing some kind of presocial nature. Yet the discursive and material power of nature serves to dampen critical debate about the social reality of greening projects. By looking like "nature," they tend to pass as apolitical, technoecological engineering projects rather than as coercive or managerial ones. When debate does arise, questions more often concern the accessibility, amenities, and/or distribution of presumed-to-be-good public resources than more fundamental questions of whether green investments are actually good at all.

I identify three characteristics of nature that contribute to these dynamics. First is nature's long-standing ideology of lying outside the social, which helps offer it up as a universally beneficial medium of social improvement in the first place. Second is the fact that greening projects occupy the space-time of leisure. Whether green spaces are designated as public or private or for individual or collective use, because they are physically and temporally separate from both work and home and their social relationships, it is generally possible to sustain an idea of these spaces as separate from social interests as well—as segregated from economic questions and forms of race, class, and gender inequality. Third is that the widespread belief in nature's universal benefit is not simply an abstraction; it is reinforced by phenomenological experiences of nature as pleasurable and as "improvement." In the Ruhr as elsewhere, exposure to leisure nature has provided a physical referent against which industrial decline has been cast as environmental improvement and economic losses reframed as environmental gains, with the fact that everyone gets an equal share of cleaner air or a new neighborhood green space a primary way in which such transformations can be interpreted as universally beneficial.

These qualities tend to foreclose public debate about the benefits conferred through greening (as opposed to more transparently "social" interventions) and to cause the power asymmetries intrinsic to greening to express themselves in the form of paternalism rather than conflict. *Paternalism* is an important descriptive, historical category in the Ruhr, referring to the region's industrial barons' labor-management practices. But, as an analytic category, greening projects are paternalistic to the extent that their proponents—whether private entrepreneurs, regional planners, or economic development officials—generally understand themselves to be providing public goods rather than offering prescriptive visions of civic life or exercising managerial power. This perception is sustained by target audiences. People know when their union is being busted or when they are being disciplined by a supervisor, but they tend to accept greening projects as investments in public goods rather than as acts of managerialism. In short, urbanized nature makes greening a useful tool for social engineering projects because its associated ideas and experiences make them appear to be about something different.

## WHAT LIES AHEAD

The book develops these themes in three parts, each consisting of a pair of chapters with their own internal contrast. Part 1 is about urbanizing imaginaries of nature and their relationship to the city as normative social and spatial ideal. It documents the emergence of urbanized nature in the Ruhr and compares the use of green space in industrial workers' housing before and after the turn of the twentieth century in order to provide a historical account of Ruhr elites' imaginative shift from *Gemeinschaft* to *Gesellschaft*—from community to urban society—and a descriptive account of urbanized nature via a contrast to the Ruhr's prior, rural-agrarian imaginary of nature. Chapter 1 shows the Ruhr's industrial barons' use of nature as a direct good for subsistence purposes in the locally unique workers' colonies (as they are called) in the second half of the nineteenth century and their imaginative turn to the city in that context. Chapter 2 documents urbanized nature's first, recognizable uses as an indirect, universal, and aspirational good in the early twentieth century by examining the region's first, flagship garden city—Krupp's Margarethenhöhe. Margarethenhöhe first established distinctions between public and private space, bourgeois ideals of urban citizenship, and a vision of the city as normative ideal in the Ruhr.

Part 2 highlights urbanized nature as an imaginary of form, not content, by contrasting two competing visions of urban democracy realized through leisure nature in the postwar period. It focuses on greening as citymaking: how the design and use of urban green space was understood to be a strategy for rebuilding urban publics and reorganizing the public realm. Chapters 3 and 4 draw on promotional materials and social science literature to show what benefits and functions two urban greening projects were understood to provide in the 1960s and 1970s, especially in relation to urban politics and public life. Chapter 3 shows how regional planners spatialized Jürgen Habermas's ideal of the bourgeois public sphere in a series of new regional recreation parks (*Revierparks*) built in the context of the crisis of democracy and the industrial economy. Chapter 4 establishes urbanized nature's availability for multiple (and counterhegemonic) visions in the same period by documenting a grassroots, New Left social movement's reimagination of the Ruhr's industrial workers' colonies as the locus of an alternative, proletarian public sphere, drawing on the work of Oskar Negt and Alexander Kluge.

Part 3 shows urbanized nature in action by documenting protagonists' and target audiences' self-perceptions of their greening acts, first from the perspective of greening protagonists, and then from that of receiving audiences. This pair of chapters uses interviews and observational data to document both the social construction of nature in action—what the imaginative and material work of greening actually consists of—and the subsequent erasure of that social organization of experience. Chapter 5 documents efforts to realize IBA Emscher Park—a new green vision for the deindustrialized Ruhr—in order to highlight the efforts in which protagonists must engage to get people to see and inhabit greening projects the way they intend despite the supposed universality of nature, and toward the ultimate end of constructing an apparently unmediated experience. With a portrait of IBA's legacy in the Ruhr city of Dortmund in the early years of the twenty-first century, Chapter 6 documents the effects of the widespread belief in greening's universal benefit by examining how greening projects are received and publicly debated. It finds that, while criticism of these projects often eventually arises, urbanized nature strongly and frequently forecloses public debate, shapes material outcomes, and enables greening as a form of well-intentioned action.

The book concludes by returning to the present, showing how the argument outlined here provides a framework for understanding the legibility of green urbanism in a wide variety of environments, and increasingly outside of the West, as this imaginary continues to travel; how these choices aggregate to transform urban environments as greening remakes cities without discussion of these practices as such; and the implications of these arguments for the politics and practice of urban greening today.

# *1*

*Green Becomes Good*

CHAPTER 1

# The Imaginative Turn to the City

Historians have traced a correlation between the rise of industrial cities and the emergence of a new, romantic view of nature as a social good in Europe and the United States in the nineteenth century. In England by the beginning of the nineteenth century, industrial capitalism had created a longing for the countryside that revealed itself in British literature (Williams 1973). In Paris, a new "metropolitan ideology" led to new forms of nature art and spectacle in the 1830s and 1840s (Green 1990). In New England, urbanization had "overwhelm[ed] the arcadian image" of Jeffersonian America and forced the development of an "urban vision" in factory towns by the 1840s (Bender 1982, 77). In France at midcentury, a "new distinctive appropriation of nature lay in the structures of the modern city" (Green 1990, 65). In New York in the 1850s, Frederick Law Olmsted and Calvert Vaux created a faux-wild landscape for "moral regeneration" with their Greensward Plan for Central Park (Crawford 1995, 64). In Chicago in the 1890s, social reformers of the Progressive movement prescribed nature as a relief from urban problems as poverty, overcrowding, and pollution intensified. In the American West, this same understanding of the contrast between the degraded city and edifying nature birthed conservation and environmental movements, including efforts to establish the first national parks, as nature was "moralized" in new ways (Farrell 2015; Taylor 2016).

Each of these is a description of urbanized nature—of nature having become a morally valued good mobilized to solve social problems and improve urban public space. In attributing this view of nature to "cities," historians have generally meant three things: the growth of slums and public health problems within cities; the retreat of wilderness and other forms of wild first nature outside them; and the emergence of a bourgeois urban culture. For centuries, so the story goes, Europeans labored to produce food, beat back encroaching forests, tame savage nature (including indigenous peoples), and cultivate wild landscapes on the continent, in colonies, and on the American frontier. As peasants and farmers were torn from the land or forced to move to cities to sell their labor, owners of means of production bought that labor, expanded industrial production, and accumulated capital in ever-larger amounts as urban poverty and immiseration grew. At this point longing for nature did too. For people comfortably distanced from wild nature and its threats, romantic ideas of nature began to look better and better in comparison to the dirt, noise, poverty, and overcrowding of industrial cities. Urban social reformers began to emphasize the benefits of exposure to city parks, fresh air, and open space for the working class, and municipal governments began to fill cities with domesticated forms of nature for both the rich and the poor. City parks, pleasure gardens, zoos, natural history museums, and animal appreciation societies all flourished in industrial cities, as did public interest in summer vacations, nature walks, wilderness conservation, and rural cemeteries.

This is the well-known story of how views of nature "urbanized," and one that the Ruhr's history of greening challenges in several ways. First, this understanding of urbanism and of nature came late to the Ruhr. Through the end of the nineteenth century, even as the Ruhr was urbanizing in the sense that it was experiencing both rapid quantitative growth and qualitative transformations in its economy and social relationships, it had not urbanized *imaginatively*, meaning that its conception of ideal society—and of nature—still reflected traditionally agrarian forms. This is visible in its industrial barons' response to a housing crisis at midcentury. Though the crisis was precipitated by industrial urbanization, another, nonurban imaginary of nature was already dominant, and for fifty years industrial barons responded by building a locally unique form of housing called *colonies* (*Kolonien*) that preserved traditionally agrarian or rural patterns of everyday life, provisioning, and social relationships in the landscape. Second, when urban imaginations of society and of nature did catch on in the Ruhr, they did so even in the absence of slums, density, and a recognizably "cosmopolitan" bourgeois culture—the main local conditions usually understood to provoke greening. It was not until the turn of the century that elites began to see the city as a social and spatial ideal to which to aspire, and, when they did, they took a popular reform model of the day—the garden city, designed as an antidote to social problems in large industrial cities—and gave it a local reinterpretation, conceiving it as a tool for forging a recognizably cosmopolitan form and culture in the Ruhr.

This chapter documents Ruhr elites' changing response to the housing crisis in order to disentangle and understand the relationships among three distinct phenomena collapsed in the industrial city origin story of urbanized nature: urbanization as a large-scale process of sociospatial transformation; what Thomas Bender (1982) calls an *urban vision* of the social world; and a moral belief in the symbolic value of nature. It provides a portrait of the colonies that serves as a "before" picture—showing what imaginaries of nature looked like before they urbanized—and identifies what conditions, if not urban density, eventually made it possible to see, talk about, think about, and represent the Ruhr and nature in this new manner. Because the Ruhr's ample green space meant that the region did not require greening in the sense usually imagined, another, crucial factor becomes visible: the extent to which the turn to the city and this new understanding of nature were *imaginative* shifts. The proximate cause of the region's urban turn was not local physical conditions but competition and a desire for influence as Ruhr elites began to see themselves as part of an international network of urban actors. As they did so, they identified the city as the desirable social and spatial form to match those ambitions and turned to urbanized nature in the form of the garden city as a means to help create it.

## THE URBANIZING RUHR

The Ruhr urbanized, in terms of the material, social, and spatial transformations that might have been expected to produce an urban view of nature, in the second half of the nineteenth century. In 1845, the region was still a quiet agricultural valley dotted with a chain of medieval market towns, small-scale mining and textile production, and a population of 300,000. By 1900, it had become a seething industrial agglomeration of over two million, one of the largest in Europe, and rivaling Germany's capital, Berlin, in size and influence. During these years, the population of the city of Essen, in the southern part of the valley on the western end of the Ruhr, grew from 4,000 to 295,000 people (Lees 1985, 1). Though explosive growth did not turn the Ruhr into a single, dense metropolis, the numbers are evidence of fundamental transformations that were taking place in social relationships and ways of life, especially a shift from subsistence living to wage labor and a market economy and to larger-scale, more heterogeneous social milieus.

As in the United States and elsewhere in Europe, new iron and steam technologies catalyzed this wave of growth by enabling natural resource extraction and industrial production on a much larger scale. As it became possible to access deeper seams of coal along the Emscher River, mines pushed north through the valley. Entirely new factory towns grew up around the mines and factories, effacing the medieval footprint laid several centuries earlier. New transportation and sanitation infrastructure—railroads, streetcars, and sewage canals—crisscrossed the region, linking the new coal mines, steel factories, and residential communities. Coal extraction jeopardized the integrity of the soil, and steel production led to flooding and air pollution. By the turn of the century, the Ruhr had become an urban-industrial agglomeration whose belching factories and expanding settlements overshadowed its old medieval cities. As one eyewitness described it in 1903:

> Virtually the entire Ruhr mining district between Oberhausen and Dortmund, Duisburg, Hattingen, and Recklinghausen already has the appearance of a continuous giant city, its individual parts linked by a thick network of electric trams, state railways, and spurs to individual mines, vibrating above and below ground with the most vigorous industrial and mining activity and teeming with a giant army of workers. It is the main artery and workshop of the German coal and iron industry. Everything lives from and for these two products and everything is touched by them. Wherever one looks there are winding towers and the broad outlines of waste-tips, chimneys, and smoking furnaces. The whole scene is enveloped and covered by a misty, gassy, dusty, dirty veil which often scarcely allows the blue sky to be seen and which falls on the rows of houses, churches etc. as a coating of dirt. (Pieper 1903 quoted in Hickey 1985, 19)

Alongside physical changes, a new set of social and economic relationships orbiting around newly wealthy factory owners and a workforce newly subjugated to wage labor appeared in this "continuous giant city." As the owners of coal mines and steel factories amassed land, wealth, and power, a new set of power brokers was produced. The coal and steel companies greatly increased in size and as an overall proportion of the region's economy during these years. In 1850, coal mining in the Ruhr was still very much a "rural" pursuit (Pounds 1968, 89–90); most of the region's 190 mines had an annual output of less than five thousand tons, employed an average of sixty-four men (usually off-season farmers), and operated in agricultural areas far from the city centers. By 1900, the number of mines remained about the same, but their average output had ballooned to 300,000 tons and their average employment to fourteen hundred full-time workers (Pounds 1968, 101). The companies that thrived were those that consolidated and vertically integrated their functions. By mining, smelting, forging, and marketing coal, iron, and steel from start to finish, they held economic power fast in the hands of an increasingly mighty few.

Workers felt the effects of these social transformations in the changing rhythms of daily life. Mining began in the Ruhr, as it did elsewhere, as small-scale, part-time work for local agricultural laborers. But between 1850 and 1900, as production and extraction capacity grew, hundreds of thousands of workers flooded into the Ruhr from the surrounding countryside and from Italy, Silesia, and Poland. This marked the transformation of the industrial labor force from local farmers who supplemented their income through off-season mining into full-time employees who supplemented wage labor with farming (Jackson 1997). It also turned them into tenants. In the early stages of industrialization, locals made money by selling off large farms and private homes for use as coalfields and building sites. They ended up renting—sometimes even the same land they had formerly owned—alongside formerly seasonal, migratory workers who had become full-time residents.

This transition from agriculture and manufacturing to large-scale industry involved a change in the employer/employee relationship from one of relative independence to one of increasing interdependence (Scott 1974). Workers were newly dependent on the factory for wages, and employers were newly dependent on workers to maintain production. As the companies grew and the production of iron and steel became more technologically and chemically precise, restless employees were an economic handicap as well as an inconvenience, and high turnover made training workers for increasingly necessary higher-skilled technical or managerial jobs impractical (Dege and Dege 1983). To cope with this situation, Ruhr industrialists adopted a "paternalistic" model of labor management (Drucker 2012), under which they provided pensions, health care, education, and housing, taking full responsibility for workers' health and happiness in exchange for their "unconditional obedience" (Günther and Prévôt 1905, 35, 181).

The firm of Friedrich Krupp AG (the germ of today's Thyssenkrupp) and the adjacent city of Essen were paradigmatic of these transformations. Essen was one of the Ruhr's medieval market towns, originally a small, walled city surrounded by farmland. Friedrich Krupp first founded a small steel factory on its outskirts in 1810, but his son Alfred turned the firm into an international weapons manufacturer that fueled German and European warmaking through the end of World War II, earning him the name the Cannon King (Manchester 1970). The company grew from 241 employees in 1850 to 68,000 in 1910, 37,002 of which were based at the original factory (Tenfelde 2005, 20). This growth transformed the town economically; it also overwhelmed it physically. As the medieval town center was dwarfed by the growing factory and adjacent residential settlements, Essen "gradually began to take on the appearance of an appendage to the factory—a symbiosis in which the guest had turned into the host." By 1900, every eighth Essener lived in Krupp company housing, and nearly 40 percent of the town's residents were directly dependent on the firm for their livelihood (Borsdorf and Schneider 2005, 127; McCreary 1968, 25, 36). By this time, Krupp's innovations in business, philanthropy, and employee management as well as metallurgy were models for other mining and steel companies not just in the Ruhr but across Germany and Europe (Günther and Prévôt 1905; Kellen 1902).

## THE HOUSING CRISIS AS AN IMAGINATIVE CRISIS

The quantitative growth and qualitative social transformations of urbanization required new visions of social organization as the agricultural valley disappeared. Though the city is not the only possible social form that could accompany industrial production, in most major industrial regions in the nineteenth century industrialization and an urban vision of society emerged more or less hand in hand. In large cities such as London, Berlin, and New York, centrality, density, heterogeneity, and the mix of political and administrative functions all helped an ideal of city life come into focus contemporaneously with industrialization as an economic form. In Paris, the "formation of a self-consciously modern and urban culture" even "*preceded* the generalized take-off of large-scale industrial capitalism" (Green 1990, 65). Berlin's designation as the state capital in 1871, a growing middle class, and increasing opportunities for urban culture, commerce, and consumption all helped its residents know they were in a city. The United States offers one historical alternative. In the nineteenth century, small-scale textile manufactures' company towns, such as Lowell, Massachusetts, initially maintained a Jeffersonian, rural vision even in the context of industrialization, imagining pastoral "factories in the forest" powered by clean water rather than dirty coal and steam (Bender 1982, 68; Crawford 1995). But, even in Lowell, by the 1840s rapid growth had made it first difficult and then impossible to maintain this bucolic ideal (Bender 1982, 75).

In manufacturing towns as well as large cities, housing was a formative site for perceptions of the pathologies and possibilities of urbanism because it was a place where the new economy and social relationships found representation in the landscape. It was, first of all, a site of deteriorating physical conditions. As Engels observed in a series of articles on "the housing question" in the early 1870s, acute housing shortages were "chronic" in cities such as Berlin, London, and Paris. In big, centralized metropolises, overcrowded, unsanitary, poorly ventilated units were the norm—from tenements in London and New York to "rental barracks" (*Mietskaserne*, so named after their military austerity) in Berlin (see Engels [1872] 1935). Housing was also a location to address these problems. Figures as diverse as Jane Addams in Chicago, Charles Dickens in London, Jacob Riis in New York, and Heinrich Zille in Berlin documented the poverty, crime, disease, and overcrowding of Victorian slums. New policy, planning, social science, and public health experts understood their task as helping decrease the bad and increase the good of modern urban life; social reformers prescribed light, air, and green space as antidotes to these conditions (Günther and Prévôt 1905; Ladd 1990, 141; Levenstein 1912; Wischermann 1986).

But, as Krupp and the rest of the Ruhr's industrial barons moved to solve a local housing crisis in the 1850s, the Ruhr's unusual morphology offered no such clear guides as to what forms of society should accompany this urban-industrial environment. As in large industrial cities, the housing crisis was a product of industrialization. But, unlike in large industrial cities, in the polycentric Ruhr housing was being built from scratch—more closely resembling housing development in manufacturing towns. Engels had also observed that, unlike older cities, which had to be retrofitted to improve social conditions, "towns which grew up from the very beginning as industrial centers"—such as Manchester or Leeds or textile manufacturing centers in the United States—were generally able to build housing faster and better keep pace with industrial growth (Engels [1872] 1935, 6). The Ruhr fit this latter profile. Housing new industrial employees required constructing entirely new residential settlements on land that lacked even basic infrastructure or services. This was true in the central part of the valley, where Krupp and other steel and mining companies housed workers adjacent to the factories, on the outskirts of the old cities, as well as in the north, where entirely new cities were growing out of the new steel and mining activity in these areas.

Also as in smaller manufacturing towns, the industrial barons were fully responsible for providing this housing and infrastructure both because it was their businesses that had produced the housing crisis in the first place and because they were in possession of most of the region's wealth and power. Germany did not become a unified nation-state until 1871. While it had begun to centralize and standardize financial institutions and infrastructure and promote large-scale investment in transportation and production technology, at the end of the century municipal governments were still barely established and received limited state support (Berger 2012; Blackbourne and Eley 1984). In the Ruhr, local governments had neither the money nor the will to solve the housing, sanitation, and transportation problems created by heavy industry (Brüggemeier 1983; Heinrichsbauer 1936; Hickey 1985, 52) and so delegated responsibility to the private sector. An 1876 law, for example, required industrial companies to provide schools, roads, water, and electricity when building housing in new areas (Hundt 1902; Larsen 1996, 975).

As a result, perhaps more than in many other places, the Ruhr's morphology gave industrialists flexibility in responding to the social crises of industrialization and in interpreting and addressing these problems. Was the Ruhr a city? Were its inhabitants urban? The landscape and people gave no clear answers. The Ruhr's lack of a division between city and hinterland meant that agrarian and industrial life remained physically intertwined in the landscape. Rather than a rural-agrarian past spatially demarcated from an urban-industrial future, fields, gardens, livestock, and woodland ran through and among its factories, railroads, cables, and canal lines and lay in the shadow of banging, hissing factories. The population was confusing too. Though the Ruhr had transformed from a region of agriculture and small-scale manufacturing into an economic force to be reckoned with, it was not a political or an administrative capital, and it had no educated bourgeoisie. Even in the 1870s, its industrial workers were still predominantly of rural origins, and "the line of division between the miner and the peasant-cultivator was not an easy one to draw" (Pounds 1968, 89–90).

In addition, the Ruhr's polycentrism meant that land remained relatively cheap because there was not one single center on which development pressure raised prices. And, because coalfields and factories required large amounts of space, as industrialists bought and built on that land they had the option to preserve to some extent the Ruhr's low-density, agricultural profile. The cost of land made this economically possible, but coal mining actually required it. The integrity of the ground was so compromised by mining activity that an 1895 ordinance actually made it illegal to build higher than four stories in Essen, and most residential units were only one to two stories tall (Niklaß 1999, 47). And so, in comparison to large industrial cities such as Berlin, the Ruhr remained uniquely rich in open space and low density throughout its industrialization (see fig. 1.1 for the local planner Robert Schmidt's rendering of this comparison).

Thus, for Krupp and the Ruhr's industrial barons, providing housing posed not just practical but imaginative questions. To what ideals should the Ruhr aspire? What form was the good life there to take? In some ways, housing construction is always an imaginative project—an opportunity to ask what ways of

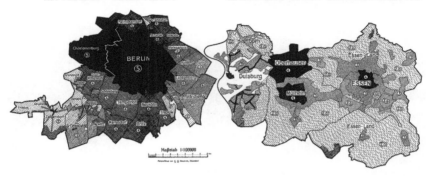

FIGURE 1.1. Comparison of housing density in Berlin and Essen, showing the Ruhr's polycentrism and greater population dispersion. (From Schmidt [1912b] 2009.)

life or normative ideals a particular community might share or cultivate—and Krupp's self-consciously paternalistic labor model strengthened this orientation by suggesting that housing was to go beyond meeting basic material needs to provide comfort and satisfaction. But in the Ruhr the answers to these questions were especially ambiguous. Though the region's social and economic relationships had definitively transformed, even though the housing crisis was produced by the new industrial economy, its affordable land and decentralized structure did not dictate housing design. Building housing required employers to envision and—through design—articulate some idea of what a good life for workers looked like in the new urban environment.

### AGRARIAN IMAGINARIES IN AN URBANIZING LANDSCAPE

Before urbanized nature reached the Ruhr, Krupp and the other industrial barons answered these questions by making ample use of another, prior imaginary of nature. Through the end of the century, they responded to the pressures of urban-industrial growth by building company housing that mimicked agricultural life. Between 1873 and 1914, they built 87,255 new homes for their workers, and, by 1910, company housing accounted for one-quarter of all rental housing in the Ruhr, as compared to 4–5 percent in "average" cities (Hickey 1985, 53; Honhart 1990). Over 70 percent of that housing took the form of a locally unique housing type (depicted in fig. 1.2), still common in

FIGURE 1.2. The Ruhr's colony housing. Note the size of the gardens and the proximity to the factory. (From Leschny-Kröger 2010, 59.)

the region today, called *colonies* (Hundt 1902, 30). The colonies reflect the fact that, though the Ruhr had, in effect, already urbanized—in terms of its physical environment, economy, social relationships, and ways of life—when called on to imagine a good life for workers, through the end of the century Ruhr industrialists still imagined a rural rather than an urban one. These small-scale communities were built adjacent to the factories and grouped workers by country of origin in clusters of semidetached houses with ample vegetable gardens and sheds for keeping livestock. Each element of the colonies' design replicated the phenomenological experiences, engagements with the landscape, and social relationships of *Gemeinschaft* rather than *Gesellschaft*: of traditionally rural, agrarian "community" rather than urban "society," even in an industrial environment.

Historians across eras and political persuasions have characterized the colonies as "village-like," and this picture began with the spatial relationship between settlement and town. Though Krupp also built dormitories and private rentals, the majority of the ninety thousand company-owned homes built in the second half of the nineteenth century were "village-style environments" with "cottage-like" housing (Wild 1983, 33) clustered near mines and factories, far from the old city centers. Proximity to the factories was partly a practical necessity—the electric tramline that would allow commuting between factory and town would not be built until 1914 (Bolz 2010, 109)—but it also reflected the assumption that a "full-time worker, who wants to care for his garden and small animals, must not be subjected to a long way" to reach them after a day in the mines (Heinrichsbauer 1936, 42). Rather than tenements or rental barracks, the brick or stone semidetached cottages, most of which had four three- to four-room apartments, each with its own entrance, mimicked the small, low-roofed, single-family homes that had already scattered the land being developed (Heinrichsbauer 1936).

The colonies were also village-like in that they created ethnic and linguistic enclaves by housing workers and their families by country of origin. Lured from Poland and the eastern provinces by the promise of jobs, foreign workers made up a third—and up to 40 percent in some areas of Essen—of the entire workforce by 1905 (Koch 1954, 20). The industrial barons intended a house with a garden, in a community with others who shared one's language and homeland, to help preserve familiar ways of life and small-scale community relationships as people "lived together in a small space and were able to continue the country ways from the rural districts from which they came, investing in a garden or plot of land that they cultivate themselves and upon which they can tend their livestock" (Pieper 1903 quoted in Koch 1954, 81). But, at the same time, these spaces were also understood to be transitional ones in which foreign workers could learn to "treasure order and cleanliness" and other context-appropriate values, such as timeliness, sobriety, and good personal hygiene, and, thus, assimilate (Hundt 1902, 31).

The most striking evidence of the colonies' rural imaginations is the relationship to the natural environment that they laid out. Rather than providing access to urbanized forms of nature—nature as an indirect, moral or affective good, as it was already being used by social reformers in large metropolises—the colonies provided a first-generation industrial workforce access to nature as a direct good in utilitarian forms: animals, gardens, and green space for subsistence purposes. An 1893 survey counted, among 16,060 miners with access to gardens and fields, 524 horses, 8,210 cattle, 31,221 goats, 38,017 pigs, and 885 sheep (Dege and Dege 1983, 50). A 1902 report found that 96 percent of all units had sheds for keeping goats, chickens, and pigs, 86.3 percent had a back garden, and 25 percent also had a small front garden (Hundt 1902, 31). Goats were so ubiquitous they were known as *miners' cows*. The gardens were attached to the residential units and included with the rental price, but, when units had no gardens or workers wanted more space, factory owners regularly subdivided the newly acquired surrounding farmland into small allotments and leased parcels cheaply (Hickey 1985). As one historian noted admiringly in the 1930s: "[An] indivisible element of almost all the company housing in the Ruhr area was and is a field or garden plot of sufficient size. In no area did workers' allotments make so early and extensive an entrance as in the Ruhr" (Heinrichsbauer 1936, 54).

The fact that these designs—small-scale, socially homogeneous communities in which one could grow one's own food—were provided not just for unskilled workers but for higher-level employees as well underscores that the colonies reflected Krupp's imagination of the good life rather than simply a (possibly poor) understanding of desires of the working class. There were, of course, practical reasons for aspects of the colonies' design. Just as agriculture and industry remained intertwined in the landscape, the transition to an industrial economy was still an incomplete one. Proximity really was important because the garden was still an extension of the male breadwinner's working day—most workers supplemented their low wages by growing food. As one worker later remarked, "There wasn't much to buy," and what food there was for purchase was often prohibitively expensive (von der Wielbecke 1965 quoted in Steinborn 1991, 5). But, even when it came to managers, for whom it was a privilege to be given a company house, Krupp offered access to gardens and animals and a factory-proximate location, just as he did the least skilled and lowest paid (Heinrichsbauer 1936, 32).

If Krupp's housing was to go beyond meeting workers' basic material needs, the design of the colonies suggests that, in those years, the good life was still understood to be a rural one. Even if tenements and urban density were not reasonable models for the Ruhr, Krupp's fantasies could have been urban and/ or futuristic. He could have constructed housing that offered elements of an urban *Gesellschaft*—placing people among strangers and offering shared public recreation areas. He could have turned to new industrial technologies and used iron and steel. Instead, through the end of the nineteenth century, he and his peers still celebrated and upheld village life as a social and spatial ideal and reproduced it in the landscape.

## AGRARIAN POLITICS AND SOCIAL RELATIONSHIPS: KRUPP'S FEUDAL PATERNALISM

The industrial barons' investment in keeping rural life and its social norms alive for their workers was a benevolent gesture, but it was one that was also constitutive of a particular type of social control. The normative ideal of agrarian community rather than urban society corresponded to a particular political and managerial vision, one in which the colonies' physical design and access to nature as a direct, subsistence good played a role in regulating behavior and subject formation. The paternalistic labor model aimed to avoid strikes and unrest partly by cultivating loyalty and gratitude rather than through direct confrontation or displays of power or control. Though desirable, the provision of company housing—along with health benefits and social services—also served important social control functions. The Ruhr's industrial barons had found it to be an extraordinarily effective solution to the problem of worker retention. Because workers had no ties to the area, housing was expensive, and wages were low, the workforce was quick to protest poor conditions or wages and quick to leave. But, in 1900, the turnover rate among those who lived in company housing was only 7.5 percent, compared to over 100 percent for those outside it (Dege and Dege 1983, 48).

The planning historian Margaret Crawford has described the paternalism characteristic of early company towns in the United States as a form of "structured dependence," in which managerial actions were linked to a "broader discourse of benevolence" and seen to provide moral, social, and physical benefits (Crawford 1995, 16–17). This was also the case in the colonies. Specifically, the relationships that the colonies established with the landscape helped evoke what is frequently described as a "feudal" form of paternalism, comparable to that of traditional landed elites, that replicated relationships between overlord and serf in new employer-employee interdependencies (Drucker 2012). The colonies were nonurban not only in their provision of subsistence-oriented forms of nature but also in their imagination of the social world: they reflected no impulse to densify, no need for great public spaces, no value in exposure to difference, no desire to cultivate an urban public life. They were equally nonurban (and nonindustrial) in their representation of power relationships. While the historical record is mute regarding exactly how explicit Krupp and the other industrial barons were about their intentions, it is clear that, in their design, the colonies' projection of rural/agrarian life encouraged a rural/agrarian understanding of politics and social relationships as well, invoking feudalism rather than industrialism and, thus, setting out behavioral expectations for both parties.

The colonies made it possible for employers to view themselves as benevolent patrons, feudal lords over new lands, while understanding workers to have "much of the essence of a peasant" (Heinrichsbauer 1936, 2), with a peasant's needs, desires, and disciplinary requirements. Potentially, this was a conflict-free alternative to seeing workers as organized, class-identified, rights-demanding citizens. To the extent that employers had become conscious of their employees as a new working class, it was mostly as a possibility to be avoided. But, if industrial workers were still understood as peasants, industrialists could imagine themselves to be more like feudal overlords than public officials or employers. If a peasant's work now simply took place below ground instead of above it, if mining was not much different than tilling soil, then the managerial and disciplinary responsibilities were clear. The industrialist-as-feudal lord's task was to keep these workers comfortable, focused, and productive by creating a domestic environment that approximated familiar (agricultural) experiences and feelings as closely as possible and disciplining them so that their behaviors were appropriate to the new labor environment, which included both personal standards of cleanliness and character and the expectation that, as a group, they need not and should not bother organizing, striking, or formulating political opinions.

Understanding industrial workers as peasant-farmers also justified keeping them in a state of docile ignorance by suggesting that education and culture were unnecessary and that (nonthreatening) activities—such as gardening—must be central to their happiness and well-being. In this way, the provision of nature was central to the colonies' managerial functions. Having a garden to care for not only created an illusion of self-sufficiency and independent provisioning; it also promised to channel workers' energy in innocuous directions. Some Ruhr workers argued, in hindsight, that the "tireless potting around in house and garden" made it more difficult to "throw a glance over the garden fence into the world" and notice injustices, get outraged, or organize a strike (Paul Bahrdt quoted in Treiber and Steinert 1980 quoted in Steinborn 1991, 12). Keeping workers in a state of ignorance was also a matter of federal policy. Because the arms produced in the Ruhr were critical to national defense and the national economy, it was important to the federal government as well as to local factory owners that workers kept working. Toward these ends, Kaiser Wilhelm II is said to have prohibited the construction of any institutions of higher education in the Ruhr. Designed to prevent contact between workers and the intellectual elite, this policy managed to prevent any universities from opening in the Ruhr until the 1960s (Eckart, Kowalke, and Mazeland 2003, 75; Keil and Wetterau 2013).

These efforts would not be totally successful. Miners' strikes broke out in the Ruhr in 1889 and 1905 and would break out again after World War I. But the social conditions created by the colonies appear to have influenced the character of local activism. Strikes in the Ruhr tended, for instance, to take the form of economically focused "spontaneous local protest," in comparison to those in nearby Düsseldorf, where more "skilled and stable" workers who were "permanently committed to urban life" had well-organized, disciplined protests closely tied to Social Democratic Party agendas (Nolan 2003, 292–93). These qualities of Ruhr protests are often chalked up to the phenomenology of mining—outcomes of men trapped together underground all day, talking about their dissatisfaction—rather than understood to be products of unrest caused by exposure to new ideas through reading, events, or organized politics (Crew 1979; see also Church and Outram 2002). But the dearth of these things in the Ruhr was in no small part a product of the colonies and corollary managerial decisions.

It is noteworthy that, even in the absence of urbanized nature, access to nature was part of a package of tools of managerial control. The presence of gardens and animals and the spatial arrangements and forms of engagement with the landscape that the colonies set out were central to establishing a particular understanding of social relationships. The colonies were facsimiles of rural life—signs of a social order that no longer existed, representations of social relationships that had largely disappeared—and the gardens and animals were the material touchstones of these activities and traditions. In suggesting that miners would be gardening after the workday and that it was important to have access to gardens and animals, the colonies contrived to make workers feel like they owned their homes even though they did not, lived in a Polish village even though they did not, and were economically independent even though they were not. In both benevolent and managerial forms, the colonies' design minimized the profound transition from subsistence to wage labor involved in the change from working above ground to working below it by implying that the miner's way of life had not changed qualitatively in the shift from an agrarian to an industrial economy.

## THE IMAGINATIVE TURN TO THE CITY

This vision of daily life and managerial relationships changed in the first decade of the twentieth century as Ruhr elites began to see the city as an ideal to which to aspire. In spite of the similarities between the Ruhr's landscape and social composition and those of smaller manufacturing regions, the ambitious industrial barons' comparative referents were never Lowell or Leeds but European capital cities. And, by the end of the nineteenth century, the Ruhr had come to rival Berlin in size and economic importance. The coal, steel, and especially arms produced in the region made it a central player in national and international economics and politics. The Krupp empire, in particular, had amassed significant money and political power. The company produced weapons to arm Germany and rival European nations and, as a result, received frequent state visits from Berlin (Manchester 1970). But Alfred Krupp's dreams of fame extended beyond national accolades and technological innovations. After half a century of scrambling to respond to the immediate needs and crises of rapid industrial growth, he began to have aspirations to social prestige as well as economic power and wanted to be known internationally for his accomplishments in social welfare and culture as well as his business sense.

The variety of public relations strategies Krupp used for his company reflected his international and urban orientation. He ran his entire operation like a stage set, curating the appearance of work and home for the benefit of national and international visitors, just as he did for his workers through the company housing. He regularly hired photographers to document the firm's growing factories and settlements, even providing strict instructions about how clear or cloudy the day should be for his smokestacks to be best silhouetted against the sky (Borsdorf and Schneider 2005; Brüggemeier and Rommelspacher 1992, 129). And he showed these photographs at every major national and international exhibition after 1876 (Bolz 2010, 97), including the Chicago World's Fair in 1893 (Beitz 1994), the General City Planning Exhibition (Allgemeine Städtebauausstellung) in Berlin in 1910, and a traveling exhibition of the German Garden City Association (Deutsche Gartenstadt-Gesellschaft) (Harris 2012, 91, 103). At these he presented the company's housing and social services alongside new weapons technologies.

By this time, a number of professionals and industry elites had begun to share Krupp's ambitions for the region. Two would be closely involved in designing the Ruhr's first garden city, Margarethenhöhe—the topic of the next chapter. One was Essen's town planner, Robert Schmidt, who was preoccupied with the Ruhr's growing urban form and particularly the role of green space in the region and laid out his urban vision in a regional plan completed in 1912 (it is from this plan that fig. 1.1 above is taken). The other was Karl Ernst Osthaus, the son of a wealthy local family that had made its money in the mining industry. Osthaus was a founding member of the German Werkbund, a predecessor to the Bauhaus, and a philanthropist concerned with the lack of art and high culture in the Ruhr (Schulte 2009). In turning to Paris, London, and Berlin as reference points, Schmidt and Osthaus, like Krupp, became increasingly aware that the Ruhr lacked the traditional spatial and social markers of industrial cities: density and cosmopolitan culture. And, in so doing, all three began to see the region's characteristic qualities—the diffuse, low-density urban form and working-class character that had been deliberately cultivated during the past half century—as problems to be fixed.

Over the course of his career, Schmidt was head of Essen's Building Commission, director of the Municipal Planning Board, a city council member, and eventually head of the Siedlungsverband Ruhrkohlenbezirk (SVR), the Ruhr's regional planning commission. His focus was on the spatial aspects of the Ruhr's urbanism. At the turn of the century, a time when it was becoming obvious that healthy urban growth required some form of planning, he was a young planner in a young profession (Von Petz 1999). Along with planners elsewhere, he saw that comprehensive, regionally scaled planning that addressed issues of green space, transportation, housing, and industrial growth in relationship to each other would be necessary—and, more specifically, that the problems of urban-industrial and rural-agrarian life could not be solved in isolation from each other (Klink and Rommelspacher 2009; Von Petz 1999). Though such an approach to planning is still unusual today, in some ways it is not surprising that Schmidt took this regional view in the Ruhr because, in fact, the local geography could not easily be understood any other way. Rather than a clearly demarcated industrial city set against an agricultural hinterland, at the turn of the century the region was (as it remains today) a collection of spatially and administratively independent settlements held together by an ever-thickening web of infrastructure.

This is to say that, while from far above or from a fast-moving train or in the shadow of a blast furnace, the Ruhr in 1900 may have had "the *appearance* of a giant city" (as noted earlier), from the ground or on foot or in the domestic life of most of its residents the sensory experience it offered was of the "misty, gassy, dusty, dirty" machinery of production rather than the social swarm of the metropolis. At the end of the century, it had retained its appearance of a chain of industrial towns and factory suburbs, clusters of developments tied only loosely to old city centers and increasingly connected by canals, railroads, and electric streetcar lines. And firsthand accounts such as Schmidt's—which describe the Ruhr in such terms through the 1960s—are notably descriptions of its industry rather than its urbanism.

Schmidt elaborated his ambivalence about these characteristics in a memorandum on regional development in the Ruhr published in 1912. On the one hand, the polycentric urban region offered distinct advantages over a traditional "great stone city with its strong, powerful center" (Kastorff-Viehmann and Utku 2012, 47). Schmidt called Essen a "leisure park" (*Vergnügungspark*) in comparison to Berlin, this by virtue of its decentralized development and low-density housing structure (Schmidt [1912b] 2009, 28–30), and argued that it was not too late to preserve and enhance the Ruhr's unique advantages through unified regional planning. But, on the other hand, the region was in chaos. Schmidt's worry was that, if it continued to grow unchecked, it would become a shapeless, undifferentiated mass—an "endless jumble of houses and trees," as Osthaus (1911a, 101) put it—into which the plague of slums would find its way just as it had in Berlin.

Osthaus was more concerned about the region's lack of cultural cosmopolitanism. In comparison to Berlin or Paris, the Ruhr also lacked a cultured bourgeoisie and the bourgeoisie's landscapes of pleasure. Industrialization had changed the region's class composition dramatically over the past fifty years. According to one study, the decline of independent craftsmen and the massive influx of unskilled factory laborers meant that a population that had been 64 percent "middle class" in 1845 was, by 1882, 75 percent "lower class" (McCreary 1964, 25), and, thanks to the lack of universities, there was little chance of education or upward mobility for this group. In addition, Ruhr cities had no equivalents to Paris's Louvre, Berlin's Staatsbibliothek, or London's Hyde Park. Public cultural amenities such as libraries, pleasure gardens, or city parks had not existed in the old medieval cities, the growing working-class population did not demand them, and the new municipal governments had not yet begun to value them—or at least not above the more pressing transportation, sanitation, and housing problems.

For Osthaus, an art historian and collector, the lack of cultural institutions in the Ruhr was more than an injustice; it was precisely because the Ruhr influenced manufacturing and produced goods that the presence of art was most necessary (Shiner 1997, 79). Exposing industrial workers to culture was essential, he believed, in order to produce better products as well as better humans. In 1902, he took steps to enrich the cultural life of the Ruhr by turning his private art collection into the Folkwang Museum in Hagen, often described as the first modern art museum in Germany. Rather than celebrating traditional paintings of German daily life and landscapes, Osthaus gathered cutting-edge work from across Europe—by painters such as van Gogh, Cézanne, and Matisse—and displayed these in combination with art and artifacts from around the world (Stonge 1993). Attendance numbers suggest that local audiences were baffled by the project; the museum received only 502 visitors in its first year (Hesse-Frielinghaus 1971, 218). Yet it must have pleased Osthaus to have provoked envy in the capital, where critics snapped that "high quality art belonged in Berlin, not in the provinces" (Schulte 2009, 215).

In the face of such absences, it was hard for those who cared about such things—most especially Karl Ernst Osthaus—not to notice a giant sucking sound coming from the general direction of Berlin. It is sometimes argued that the Ruhr is a case of industrialization without urbanization, primarily on the grounds of its lack of conspicuous consumption and culture. Yet the region's lack of urban culture was essentially a distribution problem: the wealth produced there benefited cities elsewhere. The Ruhr sweated and toiled and forged steel in its tangle of industrial infrastructure as its profits fed the cultural life of the capital, whose elite enjoyed cafés, museums, park promenades, and other forms of conspicuous leisure consumption.

## URBANIZED NATURE AS A WAY TO MAKE THE RUHR INTO A CITY

Once the city became the new ideal to which to aspire, Ruhr elites turned to nature to help fulfill it. They soon landed on an internationally available trope—the garden city—as a means through which urban form and culture could be realized. Though Alfred Krupp is said to have "dreamed of a garden city" on his deathbed in 1887 (Klapheck 1930, 31–32), the idea did not actually reach the Ruhr until after 1900. Ebenezer Howard first published *To-Morrow: A Peaceful Path to Real Reform* in 1898, founded the Garden City Association in England in 1901, and published the more widely read *Garden Cities of To-Morrow* in 1902. The German Garden City Association, which provided the institutional framework for the uptake of Howard's ideas in Germany, was founded in Berlin in 1902 (Harris 2012). After Krupp's son and successor, Friedrich Alfred Krupp, died unexpectedly in 1902, his widow, Margarethe, carried on as the family benefactor while the business scrambled to adjust. Through a foundation formed in 1906, her first act was to gift the city of Essen the land for a new housing development named, after her, Margarethenhöhe. Margarethenhöhe was intended to demonstrate philanthropic interventions as grand as Krupp's technological and metallurgical accomplishments. Both Schmidt and Osthaus were members of the influential advisory committee assembled to oversee its design, construction, and management.

The garden city is paradigmatic of the types of green solutions generated in response to the social and environmental problems of large nineteenth-century industrial cities, and it has typically been studied in those contexts. Howard initially conceived it as a fusion of town and country that offered, instead of London's social isolation, slums, and pollution, an optimal balance of nature and social opportunity, clean air and good drainage, low rents and high wages. As taken up, the idea was often understood more simply as an alternative to urban density, an escape from the industrial city and a return to nature. New garden cities on Berlin's outskirts, for example, were a weapon in "the battle against the *Mietskaserne*" (Fuchs 1908, 107). In New York, they became outer-borough alternatives to downtown tenements. Osthaus himself acknowledged this relationship between large industrial cities and garden cities in an essay about the garden city movement in which he noted: "[The concept] rises out of the hunger after light, air, movement, and good health. Social drives have called it forth; the most modern ideas of land reform have grown entangled with it. It is inseparable from the quest to put the life of the population in closer relationship to nature" (1911b, 33).

Though the Ruhr did not suffer from these conditions, the garden city solution traveled there anyway. From the beginning it was understood to solve a different problem in the Ruhr, where the local population was already in "close relationship to nature," with ample kitchen gardens and over eighty thousand animals among them. At the end of the century, the region still had relatively more affordable, better-quality housing than did large industrial cities. Though its residents struggled with pollution, flooding, and cholera, the colonies and company towns still offered, on average, higher wages, lower rents, and more comfortable housing than London or Berlin (Heinrichsbauer 1936; Schmidt [1912b] 2009) or other mining regions (Günther and Prévôt 1905; Schofer 1975). Only 80 percent of the housing was privately owned—compared to 90–95 percent in other areas (Niklaß 1999, 46–47)—and both factory housing and that built by housing cooperatives (*Baugenossenschaften*) offered better-quality units at lower-than-market rates (Helfrich 2000). Competition among the coal barons and with the private housing market had improved living conditions and lowered rents within company housing, and its presence also kept rents lower in the private housing market in the Ruhr overall (Günther and Prévôt 1905, 3; Koch 1954, 81).*

---

* This is not to say that conditions were particularly good or that there was a lack of pressing social problems. Only the luckiest workers lived in colonies, and outside and even within some of these colonies conditions remained damp, unsanitary, and overcrowded (Hundt 1902). There was never a time when new housing development could keep pace with the annual population growth (Niklaß 1999, 41), so there was a chronic housing shortage and overcrowding as people who owned their homes increasingly took in boarders or as new families shared single rooms. Air- and water-borne industrial pollution was becoming a problem, and, in addition, the Emscher River's periodic floods spread farther and farther across residential land that was slowly sinking as a result of mining-related excavation below. Factory housing built before the 1870s also lacked indoor plumbing; instead, an open sewer canal ran between the backs of the rows of houses, and there was a cholera outbreak in higher-density units in the city centers in 1866 (Dege and Dege 1983, 50; Helfrich 2000, 19). All these conditions contributed to public health problems. But it was industrial technologies, not nature, that seemed to be the right tool

Instead of the relief from urbanism it was intended to offer residents of large industrial cities, in the Ruhr the garden city found a new local interpretation—as a solution to the region's unusual morphology and class composition, used to make it more spatially and socially legible as a city. Though Krupp, Schmidt, and Osthaus all recognized the region's low-density polycentrism and its accompanying good quality of life as advantages over industrial cities, they vacillated between seeing the Ruhr as *deficient*—Berlin's gritty, provincial neighbor—and as a novelty whose unusual composition could actually be a source of an alternative urban ideal. The garden city was attractive because it could capitalize on the Ruhr's unique strengths while aligning it with more common patterns of urbanism. It offered a model of urbanism based on the interrelation of town and country—instead of urban density and cosmopolitanism—that was compatible with the Ruhr's existing social composition and spatial morphology and harmonious with the most optimistic readings of the region. In Schmidt's eyes, the promise of the Ruhr's unusual form was that it might bring the industrial town and residential settlements, as well as nature and urban society, together into one: "The idyllic valleys that emanate from the magnificent Ruhr region and that are woven deeply into its urban areas elicited the thought of whether it would be possible to unite two otherwise diametrically opposed urban forms as one, to make the industrial and residential city into a single metropolitan organism, and so to bring the most burning question of the present moment, the housing question of the metropolis, closer to its solution" (Schmidt 1912a, 34–35).

In this sense, Schmidt's understanding of the garden city was more closely aligned with Howard's initial vision than with those of social reformers who embraced the model in large cities. Like Howard, Schmidt was developing a vision of urbanism in his planning practice that, as he articulated in his 1912 memo, used nature to "unite two otherwise diametrically opposed urban forms"—the industrial and the residential city. For him, the garden city was a formula for the "schematic presentation of an incipient big city" in an

---

for improving quality of life, landscape, and development at a regional scale (Klink and Rommelspacher 2009), especially sanitation infrastructure. In addition to the expansion of railroads and electric streetcars, the Ruhr-wide water-management cooperative, the Emschergenossenschaft, was founded in 1899, beginning a century-long process of river management with the Emscher's "rationalization"—essentially turning it into a sewage canal to control flooding and improve solid waste disposal—a process that came full circle with the river's "renaturalization" in the twenty-first century, as we will see in part 3 (Peters 1999; Wuppertal Institut für Klima, Umwelt, Energie 2013).

industrial area (Schmidt [1912b] 2009, 90) built around things the Ruhr had: industry, housing, and open space. He believed that, with proper planning and guidance, the Ruhr's unique patchwork of developed areas and green open space could be cultivated and enhanced, not just to approximate a traditional metropolis, but to transcend it: "The city has to become interwoven with the countryside, and vice versa, in order to create a sound, healthy, and beautiful settlement structure, without *Mietskaserne*—a flawless organism of the modern metropolis" (Schmidt 1912a, 42; see also Von Petz 1990, 9). The garden city provided a way to accomplish this. It was possible in the Ruhr only because of its ample open space, but it could also impose order on what was threatening to become a shapeless, undifferentiated mass.

In other words, rather than a reaction against cities, in the Ruhr the garden city became a way to create them. While Schmidt was focused on the region's spatial form, for Osthaus the garden city—like his art museum—helped accomplish cityness in social terms. He saw the garden city as a means to elevate the mind, to promote the "unrestrained circulation of mental life [*geistiges Leben*]" throughout the whole organism (Osthaus 1911b, 34), believing that, like exposure to art, exposure to the city provided "salvation from the crudeness of daily life" and was a route toward more enlightened, sensitive, sensuous ways of living in industrial modernity (Umbach 2009, 101). In an essay titled "The Meaning of the Garden City Movement for the Artistic Development of Our Time," he wrote: "I imagine the world city of the future as a heart of concentrated energies, out of whose interaction the culture of the future grows, and around this core a radial system of residential areas—garden cities, if you like—between which wide promenades, sport fields, and playgrounds gradually expand to field, forest, and meadow" (Osthaus 1911a, 99).

Osthaus saw a close link between the garden city as a spatial form and the "cosmopolitan" culture that was supposed to grow from it. Elsewhere, he railed against interpretations of picturesque garden cities as receptacles for nostalgic or agrarian ideals. It "seems absurd to me," he declared, that the garden city form should be mistaken for a small-town spirit. It was a "child of the modern metropolis" and, as such, "should be, not a fall back into the half culture of the small town, but an outgrowth of the metropolis" (1911b, 33, 38). The garden city as a retreat *to* nature *from* the city and its culture, enlightenment, and progress—whether expressed in architecture as pious historicism or in planning as romantic nostalgia—was a withering, a "withdrawal" from big-city culture that reflected only the parochialism of an "overstuffed petit bourgeoisie [*gesättigten Kleinbügerlichkeit*]" that was as socially regressive as it was spatially incoherent (Osthaus 1911b, 38, 34). For Osthaus, the garden city was simultaneously an expression of the cosmopolitan culture of the future and a way to create it. Though he was less elegiac than Schmidt about the Ruhr's present, like Schmidt he believed that the garden city could help the Ruhr develop an urban form that could transcend the European metropolis. And both wrote in a tone of collective participation in a great unfolding, in which the garden city promised to fulfill their ambitions of an urban future. For all three—Schmidt, Osthaus, and Krupp—the garden city appeared to be a way to produce cities and citizens in the Ruhr: to turn an amorphous, provincial region into something with a recognizable city form and culture.

## INTO THE URBAN FUTURE

The garden city marked the arrival of urbanized nature in the Ruhr as, once the city became thinkable, greening became thinkable as well. The main historical insights of this chapter—made visible by the Ruhr's morphology—are that the region's industrial barons were able to maintain an agrarian ideal of society for half a century even in the context of rapid urbanization, an ideal that they reinforced in the landscape in the form of subsistence agriculture and small-scale, homogeneous communities, and that the catalyst for local elites' imaginative shift to an urban future was their coming to see themselves as part of a new reference group within which these concepts were imaginatively available. Analytically, this history begins to highlight—even before the adoption of urbanized nature—how the built environment was shaped by these imaginative choices as well as structural factors. Urbanization set the terms of development and guided the parameters of change within which the imaginative turn to the city could take place by setting in motion huge political, economic, geographic, and social changes, by creating a new class of industrial elites in Europe and the United States connected in material terms, and by producing, by the end of the nineteenth century, large urban agglomerations containing both positive and negative aspects of urban life. But, in a place where it was physically possible to construct relationships between people and the landscape characteristic of an agricultural economy even in the context of industrial production, an urban vision was neither a necessary nor an obvious choice, and agrarian imaginaries initially prevailed in spite of those objective transformations.

The Ruhr's turn to the city as a normative ideal and to greening as a vehicle for these sentiments, then, complicates the industrial city origin story of greening in several ways. It suggests that the local condition of possibility for greening is not a particular (dense, pathological) urban morphology: in the Ruhr, a lack of green space and residents' apparent alienation from nature were not required for garden cities to be thought of as a useful solution to very different urban problems. It establishes that an urban vision of the social world is not an immediate and necessary consequence of urbanization, understood as a set of widespread material transformations of the economy, social relationships, and the landscape: agrarian imaginaries thrived in the Ruhr even in a transforming environment. And it identifies the critical role of the imagination in establishing a new, public urban vision: in the Ruhr this crystallized among the region's elite fifty years after major transformations in the physical environment and social and economic relationships had already taken place.

Sociologists understand major transformations in political and economic structures and in ways of life—such as those brought about by industrial urbanization—as "unsettled" times when cultural logics and codes of civic life come up for debate and new repertoires of social action can emerge (Sewell 1996; Swidler 1986). New concepts and forms of action can be a direct result of local events or activities—such as greening as a reaction to experiences of dense, degraded industrial cities—or influenced by external ones, at different scales. Benedict Anderson, for example, argued that a crucial factor in the abstract idea of a nation becoming meaningful in a particular place was local actors coming to understand themselves as part of a "competitive, *comparative field*" defined at least in part by that concept (Anderson 2006, 17). This latter situation was the case in the Ruhr. The region's industrial barons' initial response to local transformations was to seek recourse in representations of rural-agrarian life. But another outcome of industrial urbanization was local elites becoming part of an international field of other cities and urban actors. And, fifty years later, even with the colonies reinforcing rural-agrarian norms in the landscape, they made an imaginative turn to the city as they were influenced by this new reference group. It was not local conditions or material/social transformations that were responsible for Ruhr's urban turn but elites' shift in comparative referents and social aspirations—coming to see themselves as part of this global field that was itself a product of urbanization.

Having seen the Ruhr's preurbanized social norms and use of nature in this chapter readies us to consider what is unique about urbanized nature in the coming one. Nature in the Ruhr in the nineteenth century was still considered "good" and still had managerial functions; treating it as a subsistence good did not make it any less desirable or any less a mechanism of control. But, after imaginaries of nature and ideals of society urbanized, nature's perceived goods and managerial functions changed—from subsistence goods creating an illusion of material self-sufficiency to moral or affective goods supporting conceptions of individual freedom and the cultivation of a bourgeois self. Chapter 2 documents urbanized nature's aspirational use—as a bearer of indirect, social or moral goods—through the construction of garden cities in the Ruhr at the beginning of the twentieth century. Not only were these changes in imaginaries and uses of nature evident in Krupp's Margarethenhöhe, but they also echoed throughout the region.

CHAPTER 2

# Building an Urban Future through Nature

In 1989, the geographer Margaret FitzSimmons entreated urbanists to see "Nature" as a category constructed through "the urbanization of consciousness," much like "History, Geography, and Space." "We see Nature," she argued, "through the geographical and historical experience of the urban": "Nature as we know it was invented in the differentiation of city and countryside, in the differentiation of mental and manual labor, and in the abstraction of contemporary culture and consciousness from the necessary productive social work of material life" (FitzSimmons 1989, 110, 108). That, in essence, is the transformation this book elaborates and the social consequences it documents—what FitzSimmons describes very broadly, as the production of nature in the First World, but which I see more narrowly, as a historical transition that made it possible to green cities, first in Europe and North America but increasingly across the world, in the familiar, contemporary sense.

In the Ruhr, this new, urbanized conception of nature was first put to work, materially, with the construction of garden cities in the first decade of the twentieth century. Historians of the region do not generally describe garden cities there as a dramatic break from the colonies that preceded them. As Osthaus acknowledged when he expressed frustration that the garden city could be mistaken for representing "small-town" culture rather than being properly seen as a "child of the modern metropolis" (1911b, 33, 38), in purely aesthetic terms it is easy to see garden cities as continuous with colony ideals, expressing romantic nostalgia for agrarian life rather than a modern, urban vision of society. In Essen, Margarethenhöhe's architect, Georg Metzendorf, aimed to create a "magic little town" (Hall 2002, 115) separated from the city by a wooded greenbelt and made accessible by a small stone bridge. Margarethenhöhe followed the style of German garden cities, which were particularly ornate thanks to the influence of the Austrian architect and planning theorist Camillo Sitte, whose 1889 *City Planning according to Artistic Principles* celebrated the organic forms and curving streets of medieval cities (Collins and Collins 1965). Crossing the bridge into the settlement today, as in 1910, you find a central market square—with buildings that were originally the company store and guesthouse now a supermarket and a hotel-restaurant—that is surrounded by narrow, curving streets lined with gabled facades, shuttered windows, picket fences, and climbing ivy. Garden cities and colonies are also formally similar: both are models of small-scale residential settlements that place nature at the center of the good life, provide each unit its own open space, and are set apart from the outside world.

Moving beyond their aesthetic similarities, viewing the Ruhr's garden cities as nostalgic expressions of agrarian community rather than urban society is also consistent with prior historical interpretations: as a re-creation of a fairy-tale German village in a landscape brutalized by industrial production (Gesellschaft für Heimkultur and Hecker 1917) or, under National Socialism, a *Heimat* or national landscape worthy of preservation (Heinrichsbauer 1936). And at the beginning of the twentieth century the garden city did, in fact, arrive in Germany along two routes almost simultaneously—one (Howard's) progressive and reform minded, the other reactionary. Two years before *Garden Cities of To-Morrow* was published, and nine years before it was translated into German, Theodor Fritsch outlined an anti-Semitic, ethnonationalist version of the garden city in his 1896 *Die Stadt der Zukunft* (The city of the future). Fritsch's garden city would, in fact, eventually be celebrated by National Socialists (Schubert 2004), but, two decades before its appropriation by the Nazis, it was part of a family of cultural, environmental, and lifestyle movements emerging across Germany that have been traditionally interpreted as reactions against industrial modernity.

But, in spite of Margarethenhöhe's aesthetic similarities to the colonies and its possible resonances with nationalistic and antiurban agendas, I argue that the garden city marked a decisive break from the colonies in that it reflected a distinctly urban vision of both society and nature. It was also the first of many recurrent efforts to turn this working-class urban region into a legibly middle-class city. Empirically, its urban vision can be confirmed by its founders' affiliations with international social reformers and ideas. The reactionary Fritsch never mentioned Margarethenhöhe in his writing on the uptake of his ideas in Germany (Schubert 2004), but the settlement was promoted by the

German Garden City Association, which saw it as being in alignment with its notion of the garden city's formal and social goals (Harris 2012, 5). In addition, as chapter 1 argued, Schmidt, Osthaus, and Krupp—each ambitious and internationally oriented—looked not to other Germans but to other industrial cities for inspiration.

Theoretically, this new orientation to the city is also visible in Margarethenhöhe's use of nature as a space of leisure rather than labor—what FitzSimmons describes as nature's "abstraction . . . from the necessary productive social work of material life." While the colonies provided access to nature for subsistence purposes, at Margarethenhöhe animal keeping and subsistence agriculture were expressly forbidden. Instead, gardens and green space were used in the contemporary sense—to fulfill an urban-aspirational rather than a rural-preservationist vision of society and understood to deliver indirect, moral and affective goods rather than direct, material ones. This chapter uses the case of Margarethenhöhe to establish a description of urbanized nature in a paradigmatic form. It also begins to illustrate greening's social and spatial consequences as these new perceptions of nature found affinities with new liberal forms of managerial power and control.

## THE LEGIBILITY OF NATURE AS AN INDIRECT GOOD

Urbanized nature is marked by the consistent construction of signifiers of nature as indirect, aspirational, and universal goods. The first and most obvious change from the colonies to Margarethenhöhe is in nature's goods from *direct to indirect*—from the provision of nature as a food source to something morally and spiritually beneficial. This new orientation is reflected in Margarethenhöhe's leases and house rules, which are on view today in a small visitor's center in the garden city.

While animals proliferated in colonies through the end of the nineteenth century, Margarethenhöhe's house rules state: "The keeping of rabbits, chickens and free-flying pigeons is forbidden" (Margarethe Krupp-Stiftung für Wohnungsfürsorge 1915). In addition, though each unit still had its own green space, it was not to be used for growing food. Unlike the colonies, which had expansive kitchen gardens, Margarethenhöhe offered residents tidy green yards, each planted with one decorative fruit tree and separated from its neighbors by shrubs and greenery (Metzendorf and Mikuscheit 1997, 32–34). And, though Alfred Krupp had once found it both "economically (i.e., livelihood contributing) and morally very useful for working families to have a garden culture" (Günter 1970, 154), by 1887 he registered concern that, "if the garden . . .

is not controlled by the woman alone, then [the husband] works at home and rests at the factory." Thus, he declared, "subsistence agriculture must not be practiced" (Günter 1970, 154).

These very different uses of nature at Margarethenhöhe mark the moment that green "became good" in the Ruhr. The shift from the colonies to garden cities was a shift from providing access to nature for subsistence purposes to actually forbidding such uses. The garden cities replaced utilitarian and materially necessary forms of nature with essentially decorative, symbolic forms: nature as an indirect good, morally charged as positive. Of course, as nature's subsistence value in the colonies highlighted, agrarian imaginaries of nature also rendered gardens and animals as "good," with corresponding moral connotations. And this imaginary of nature continued—along with others, such as a vengeful and/or forgiving wild nature one might pray to or a utilitarian, useful nature-as-resource to be exploited—to coexist with the emergent, more contemporary conception. But Margarethenhöhe's use of parks, trees, animals, and green space illustrates the emergence of urbanized nature, our object of analysis, in the Ruhr.

I have said that this view of nature is often described as urban: as a characteristic of "city people" (Nash 2014; Schmitt 1990), as "an urban image of the country" (Bender 1982, 92). If the Ruhr lacked both traditional cities and traditional city people, how did urbanization produce this view of nature? As chapter 1 argued, urbanization contributed to the spread of this idea among elites—who had had access to pleasureable forms of nature for leisure and recreation for much longer—by situating the Ruhr's industrial barons in a new comparative, competitive context: a global urban field where social reformers in large industrial cities were already "improving" cities by greening them. But, in addition, urbanization fundamentally transformed experiences of nature at the quotidian level, producing new cross-class, everyday experiences of nature that helped make these ideas widespread and widely legible for the first time. These changes were also described in chapter 1: a shift from working the land to working in factories, from subsistence existence to selling labor, from making food and clothes to buying them, and from a context of relative social homogeneity and stability to one of far greater heterogeneity and mobility. This chapter demonstrates that one outcome of these changes was to turn nature from a space of labor into one of leisure. As the working classes shifted from agricultural to industrial labor and from subsistence to a market economy, recreational engagement with nature became increasingly available. Through these changes, gardening became a hobby for women and children instead of work for men, and the garden came to be commonly understood as a space of relaxation and recuperation—experiences that made it possible for working people to relate to a view of nature as scenic.

Upton Sinclair's 1906 muckraking novel *The Jungle*, about immigrant labor in Chicago's meat industry, provides an excellent portrait of then-new ideas of nature as a moral leisure space as they appeared to reading publics at the end of the nineteenth century. After many hardships, Sinclair's protagonist, Jurgis, escapes Chicago and its slaughterhouses for the countryside, where,

> [w]henever the cars stopped a warm breeze blew upon him, a breeze laden with the perfume of fresh fields, of honeysuckle and clover. He snuffed it, and it made his heart beat wildly—he was out in the country again! . . . [F]or three long years he had never seen a country sight nor heard a country sound! Excepting for that one walk when he left jail, when he was too much worried to notice anything, and for a few times that he had rested in the city parks in the winter time when he was out of work, he had literally never seen a tree! And now he felt like a bird lifted up and borne away upon a gale; he stopped and stared at each new sight of wonder,—at a herd of cows, and a meadow full of daisies, at hedgerows set thick with June roses, at little birds singing in the trees. (Sinclair [1906] 2006, 235)

Jurgis's elation and these descriptions of nature (familiar to us, of course, because these are the same sentiments motivating greening practices today) reflect urbanized nature's connections to these structural transformations. Nineteenth-century industrial workers like Jurgis were part of the first generation of Europeans to experience natural landscapes as something other than spaces of hard labor—fields that needed tending, cows that needed feeding—or potentially deadly, wild threats, in the Old World or the New. Though he had moved to the city to sell his labor only three years earlier, and in spite of the fact that his work at the slaughterhouse included ample contact with nature being processed on a larger scale (he and his fellows were butchering cows, much as Ruhr coal miners were still engaged in natural resource extraction), Jurgis could now view agrarian life—cows, daisies, and hedgerows—with "wonder." He also had a new economic relationship to the countryside. Following this moment of nature euphoria, he has a confusing encounter with a farmer. He asks for breakfast. The disgruntled farmer thinks Jurgis is a tramp looking for a handout. But it is a misunderstanding—Jurgis was expecting to purchase this meal (and, later, a bed) using his wages from the city. After he eats, he declines the farmer's offer of work and lies in the shade of a bush, on a riverbank, "for hours, just gazing and drinking in joy." What the farmer views, suspiciously, as laziness is simply the wage-earning city dweller's experience of the countryside: Jurgis is enacting new, urban associations with nature as a moral leisure space. His leisure and joy in the working countryside is the mark of his having become an urbanite and the miscommunication with the farmer a measure of the gap between the two worlds.

Jurgis's euphoria—like many popular examples—also reflects the way we usually perceive nature as good in contrast to a concrete jungle: "For three long years . . . [Jurgis] had literally never seen a tree." But, again, this experience of urbanism never fit the Ruhr, where colony residents kept thousands of pigs and cows, tended vegetables after the workday ended, and lived in small clusters of residential dwellings surrounded by green space that in many ways prefigured Howard's garden city ideal. The construction of garden cities in the Ruhr, then, encourages a different reading of the contrast between urban society and country nature in *The Jungle* as well as a different set of ideas about the conditions under which this view of nature might be legible. While the spatial and experiential contrast between town and country—the dense industrial city versus the pastoral countryside—might be one source of these perceptions and associations, the Ruhr demonstrates that urbanized nature was legible within this set of broader transformations that transcended cities—from agriculture to industry, from subsistence living to wage labor—even without such local contrasts. Just as Jurgis's ongoing experience with nature in the form of meat processing did not inhibit his formation of a romantic view of the countryside, the Ruhr's local abundance of signifiers of nature did not prohibit Krupp or Osthaus from adopting urbanized nature either. For people embedded in urban-industrial social relationships, and incorporated into a modern industrial labor force, it was possible to experience a country idyll as morally and socially beneficial.

## GREENING AS CITYMAKING

Urbanized nature's second characteristic—its *aspirational* quality—signals both the future-oriented and the normative aspects of greening practices. From the beginning, greening was a means of achieving, rather than an escape from, urban life, and nowhere were the Ruhr's new urban ambitions and urbanized view of nature more fully realized or easier to see than at Margarethenhöhe. Margarethenhöhe was a public relations project—extraordinarily well resourced, reflexive about its intentions, and self-consciously positioned on a world stage. After two Krupp patriarchs had died in quick succession, the widow Margarethe's first act was, as we have seen, to make a gift of land to the city of Essen for a new housing development intended to extend the firm's investments in the city and showcase Krupp's and the Ruhr's successes for an international audience. There, the Krupp family and the architect Georg Metzendorf used the same design elements as were found in the colonies—gardens and green space—to construct the region's first urban public spaces and its first urban public and forge the social experiences and social consciousness of a modern urban middle class, thereby turning a diffuse, working-class urban region into something socially and spatially legible as a city.

## Fiefdom to Suburb

The first thing the Krupp family did was elevate the city of Essen by defining Margarethenhöhe relationally, as a "residential enclave of greater Essen" (Kallen 1984, 48). Throughout the nineteenth century, the mines, factories, and workers' colonies had grown and operated independently of and remained indifferent to the Ruhr's medieval cities. They were far enough away that residents of company housing rarely, if ever, spent time in the old city centers (Günter 1970, 151; Kellen 1902, 171). But, by 1900, the Krupp steelworks was no longer an isolated factory in the hinterlands, but rather a sprawling complex whose increasing "gigantism" had begun to dwarf the medieval city of Essen (Borsdorf and Schneider 2005, 127). Margarethenhöhe inverted this relationship by bolstering the authority of the city to which it was a satellite. Rather than functioning as a fully independent garden city in keeping with Howard's model or an autonomous fiefdom like the colonies, it was to be a suburb, part of but separate from the city in its location, daily rhythms, and administration.

An electric streetcar system introduced in the 1890s first made it possible for residents of Margarethenhöhe to commute to work in the city (Jackson 1997, 139; McCreary 1964, 189), accessing all the "advantages of the metropolis" while maintaining a pleasant separation of functions (Brandi 1912, 43–44). This relationship was also built into the site operationally and aesthetically. Metzendorf's 1906 project memorandum begins by describing the settlement as an "artistic, unique district of Essen affiliated with the city center" (Metzendorf 1906, 4). In addition, Margarethe Krupp donated fifty hectares of adjacent woodlands and creek to shield the settlement from the bustle of the city and the noise and pollution of industrial production (Steinhauer 1956, 37). This greenbelt was a transition zone, allowing the settlement to, as Metzendorf put it, "freely and naturally give way" to the woods and cottages of the surrounding countryside. Metzendorf also described the approach from Essen in detail: the experience of seeing Margarethenhöhe from a distance, of crossing the creek that runs along the front of the property via a bridge that both "linked the city and the settlement" and acted as a threshold to "a world apart" (Metzendorf 1906, 2, 6).

Though the factories had been the actual engine of economic growth in the preceding decades and the coal and steel barons the real locus of economic and political power (it was, for example, Margarethe Krupp who had the privilege of introducing the mayor of Essen to Kaiser Wilhelm II, not the other way around [Helfrich 2000, 91]), Krupp increased Essen's municipal authority with the establishment of Margarethenhöhe. Unlike the colonies, where the company had retained full control, the Krupp family explicitly gifted Margarethenhöhe to Essen and surrendered much of the garden city's administration and management to the city (Helfrich 2000). They made Essen's mayor chairman of the foundation (as he remains today) and included local and regional administrative officials, along with Robert Schmidt and Karl Ernst Osthaus, on the advisory committee formed to oversee the design. The city even had the deciding vote on the committee (Steinhauer 1956, 33). Margarethenhöhe was also the first company-built residential community to serve the public, offering housing to those not employed by the firm, and designating the surrounding woodlands and creek a public park (Krupp'sche Gussstahlfabrik 1912a, 314).

## Labor to Leisure

New urban ideals were especially apparent in Margarethenhöhe's rules and regulations for engagement with green space. The prohibitions against agriculture and animal husbandry enforced new, leisure experiences of nature and also aligned with several other recommendations for improvements to the old colony design that Alfred Krupp had made before his death: that modern amenities such as schools and indoor toilets should be prioritized over sheds for animals and that, when garden spaces and sheds were provided, they should be of a scale appropriate to their status as accessory, recreational spaces for women and children, rather than necessities (Klapheck 1930).

With these changes, the garden became a private, family space—a place to recover from the workday instead of an extension of it, and the domain of women and children rather than men. In Metzendorf's words: "[Its new value was] as a space of recuperation and work space at the end of the day's duty, that shouldn't be a—usually badly designed—showpiece for passersby on the street, but should be, in good seasons, an extension of the living space. It [the garden] fulfills its true inner purpose only when it serves the social life of the family. No unbidden stranger, therefore, has a right to glimpse inside!"

(Metzendorf 1906, 12). In defining the garden as an extension of the home and the family, Margarethenhöhe's developers imagined wages high enough and cheap goods plentiful enough that families need not grow their own food. They imagined a physical separation between the realms of production and social reproduction that restricted the male breadwinner's work to the factory. They assumed a division between public and private spheres in which streets could be imagined to contain both strangers and passersby from whom one might want (and could be expected to obtain!) some privacy in one's garden. And they imagined a new role for the garden itself, as fulfilling spiritual rather than material needs.

This new relationship to nature as leisure extended beyond private gardens to public green space. As backyard gardens turned from kitchens into living rooms, the surrounding agricultural land was turning from working farms into a site of café culture. Margarethenhöhe's adjacent woodlands were to serve as a *Naherholungsgebiet* (Kallen 1984, 55), a term that might be not so succinctly translated as "a nearby area for recuperation and rejuvenation"; the idea was that residents might spend weekday evenings strolling along Margarethenhöhe's creek enjoying the songs of the nightingales (Klösters 1982, 196). As the settlement expanded around one of the remaining farms in the region, in 1904 Metzendorf helped the owner reconceptualize his failing agricultural business as a commercial leisure space—a "garden-pub and excursion destination" (Metzendorf and Mikuscheit 1997, 67–68; see fig. 2.1). This produced new ways to interact with the environment that transcended the housing itself, such as the opportunity to enjoy a farm as a location for cake and coffee, being warmed by a stove decorated with animals instead of working among them.

## Peasant to Bürger

Creating citizens required new forms of exposure as well as private retreat. The colonies had grouped residents by country of origin, skill, and salary level to protect them from new things and people. A 1902 newspaper article had complained that the colonies were "Polish enclaves on German soil," while a colony resident remarked in 1913: "For the children [the colony] is worthless. They neither see nor hear anything better and the bad example makes them forget all my exhortations" (Fischer-Eckert 1913 quoted in Hickey 1985, 66). Now, to produce people accustomed to diverse interactional environments who had an appropriate regard for distinctions between public and private space, decorum in public, and an appreciation of urban culture and amenities, Margarethenhöhe's design elevated diversity and cosmopolitanism above

FIGURE 2.1. Margarethenhöhe's new relationship to nature and café culture. (From Metzendorf and Mikuscheit 1997, 67.)

familiarity and homogeneity, emphasized cultivating the mind over fortifying the body, and taught discerning consumption instead of providing for individual production.

The decision to make Margarethenhöhe a socially mixed settlement that offered housing to noncompany residents as well as "Kruppianer" for the first time supported the company's socialization goals. The suburb was intended to be for the lower classes (*minderbemittelte Klassen*) (Brandi 1912, 43), though in reality only longtime (and therefore relatively highly skilled and well-paid) Krupp employees managed to live there. Of the non-Krupp renters, who made up about half the suburb's residents in 1913, most were public-sector employees of the "new middle class" (*neuer Mittelstand*) who worked in administration, law enforcement, or education, or for the postal service, the railroad, or the courts (Bolz 2010; Kallen 1984, 73). Making neighbors of such disparate types of people was ostensibly to "cultivate true citizenship among all elements of the industrial city" (Steinhauer 1956, 33).

Richard Sennett has remarked that cosmopolitanism was "passed to the working classes as an experience of consumption" (1977, 137). And, indeed, as Krupp men were being conditioned to wage labor in the factory, in Margarethenhöhe Krupp women were being trained to create and maintain a pleasant middle-class home. By 1900, Krupp's company stores had expanded to sell more food and home goods more cheaply, and wages had increased. The company's educational programs included domestic arts classes that taught women how to stretch their husbands' wages and improve the efficiency of their households (Hickey 1985, 58; McCreary 1964). And Margarethenhöhe's residential units were designed with these new citizens in mind. The single-family homes accommodated no more than three children per couple and featured indoor plumbing, separate baths and kitchens, more efficient and comfortable heating systems, and amenities such as wallpaper and furnishings that had until recently been unavailable to the working class. House rules also forbade a variety of formerly common activities that blurred distinctions between public and private space or spheres of production and social reproduction—such as taking in boarders or selling things from the home—and specified aspects of domestic behavior such as hours during which carpets and laundry could be hung outside to dry (Margarethe Krupp-Stiftung für Wohnungsfürsorge 1915).

All these changes were part of a deliberate embourgeoisement of the Ruhr's working class and reflected a changing view of industrial workers, who had formerly been understood as peasants, into being seen as *Bürgers*—middle-class members of an industrial economy. The question of the *Bürger*'s definition and identity has inspired vigorous scholarly debate (Sperber 1997), and, if considered in terms of economic status, this group was a real minority in the Ruhr. But, as a new subject type, the *Bürger* is recognizable by his (for he was usually a "he") norms of cosmopolitan citizenship: a sense of individualism, a positive attitude toward work, a civic orientation, and a respect for culture and scholarship. Margarethenhöhe's physical design, rules, and social composition all helped cultivate this subjectivity in the Ruhr's industrial workers and established patterns of private and public life that helped instutitionalize a public sphere (Calhoun 1992, 10) for them to inhabit in the Ruhr. And, while there were national inflections of these ideals, in terms of both ethnocultural and institutional affiliations with the new German nation-state (Brubaker 1992), locally this new subject was concretely identified with the historical form of the city.

## GREEN BECOMES A PUBLIC GOOD IN ESSEN

The third characteristic of urbanized nature is its perceived *universality*: it is understood to be beneficial to all and to all in the same way. Again, this classic conception of an a- or presocial nature is premised on long-standing society/nature binaries that suggest that nature is something that all humans—regardless of class, culture, or historical experience—will have the same relationship to. The construction of nature as asocial also gives it special properties in social contexts. Imagined as a "moral preserve" that is "free from social interests" (Bell 1995, 138, citing Tavory and Jerolmack 2014, 70), it can be seen as a blank slate or an empty vessel with an unproblematic and uninterested relationship to human arrangements.

This old idea received a new meaning in the context of urbanized nature. Before urbanized nature, animals and gardens were provided for all but remained matters of the private realm—part of the reproduction of family life and the provisioning of the household. Once nature became indirect good, however, it also became a *public* good, a matter of the public realm (now that there was a public realm it could be a matter of), which is to say that it received public funding and functioned as a public space. This was the case with Krupp's philanthropic acts; with Margarethenhöhe the company treated the provision of nature as a public good for all citizens of the city. In order to reasonably attribute greening in the Ruhr to a change in the social imaginary, however, we should see greening being taken up more broadly, and, indeed, these changing sentiments were visible not just at Margarethenhöhe but throughout the region over half a century.

In the public sector, they took the form of new attitudes toward what had been called, derisively, "decorative" forms of green space (Von Petz 1990). Large European cities had first begun to provide flower beds, street trees, and parks alongside cultural institutions, large-scale infrastructure, and other amenities for a growing leisure class in the eighteenth century (Ladd 1990). Throughout the nineteenth century, as these things came to be seen as important for all city residents rather than simply elites, this use of nature diffused spatially as well: parks, tree-lined boulevards, and public buildings spread from the city centers, where the wealthy and middle class lived, to working-class suburbs (Ladd 1990, 214).

In the Ruhr, this process did not begin until the mid-nineteenth century and moved in the opposite direction: urbanized nature began in factory suburbs and was led by private investment. The industrial barons, for instance, had used tree-lined streets and flower beds to relieve "monotony" in company housing as early as the 1870s (Dege and Dege 1983, 52; Heinrichsbauer 1936, 28), long before the city of Essen could see street trees as anything other than a frivolity. As mentioned in chapter 1, this was likely partially a consequence of the fact that, for young municipal governments, building parks and plazas was simply less critical than solving planning, sanitation, solid waste disposal, and transportation problems. But, regardless, from about the 1860s to the 1920s, municipal government action reveals a transforming sensibility spreading in the Ruhr as city governments began adopting greening modes characteristic of large industrial cities. Their willingness to take over these responsibilities from the private sector suggests a change from seeing trees and parks as luxuries provided by private dollars to seeing them as public goods to be provided alongside other municipal amenities.

## Green Space as Urban Culture

Growing municipal support for greening required a reorientation toward green and its role and not just a reallocation of time and resources. Perhaps because there was such a strong historical correspondence between decorative greening and the presence of a wealthy leisure class in nineteenth-century cities, nature appreciation has often been seen as a property of this class—for example, the urban elites who led conservation and nature-oriented social reform movements (Taylor 2016). This was precisely the class that was missing in the Ruhr. Ursula Von Petz, a German planning historian, has written of the region: "Conventional measures of park (i.e. 'green open space') design that had become fashionable in bourgeois metropolitan cities, sparked little interest in the working class district. . . . [H]ere, people found it difficult to pay attention to matters of prestige; here, the *flâneur* or the promenading members of the *haute bourgeoisie* could not be found. Here, one toiled; the workers—in their housing settlements—were granted plots for subsistence gardening which were soon covered by the soot of furnaces" (Von Petz 1999, 168). But Krupp, Osthaus, Schmidt, and local business leaders aimed to create public life and an urban bourgeoisie by providing nature and civic space. For these entrepreneurs, greening the Ruhr's cities was, as at Margarethenhöhe, less a means of meeting an existing leisure class's demand for nature than it was a means to create that leisure class. It also reflected the increasingly widespread view of nature as a public good for all classes that, by the nineteenth century, had become common across large cities.

Greening in Essen began with the creation of large green spaces that could serve as a public realm. The first were products of private money and aimed to create conditions for promenading and public display—part of constructing, through green space, an "infrastructure of metropolitan life" (Kaschuba 1995, 109) that already existed in cities like Berlin. Like Margarethenhöhe's privately funded public woodland, an association of private businessmen established Essen's first city park, the Volksgarten, or "people's park," in 1863. The park contained a public building with an open terrace for performances, events, and *flânerie* and served as a place to perform class location and engage in conspicuous consumption (Horst 1994, 42).

The city administration gradually took on responsibility for building and maintaining parks and green spaces as they came to be perceived as public goods in Essen. In 1881, the Volksgarten was renamed the Stadtgarten (the contemporary word for a "city park"), and, in 1888, the city assumed financial responsibility for it, enlarging it at the same time (Horst 1994, 14, 42). As Essen's population and pollution continued to grow, in 1902 the city decided to build a second city park, the Nordpark, that also contained a public building. Around the same time, the city council voted to take care of a row of privately donated trees along the Huyssens Allee and plant trees on a parallel street; the "relatively insignificant" amount of money involved may nevertheless have reflected a "changing attitude" toward these spaces (McCreary 1964, 127). In 1904, after ten years of advocacy, the reluctant city council finally agreed to purchase the municipal forest, and, in 1907, it purchased five hundred more acres to expand the area. In 1914, the city also began building smaller neighborhood parks to serve as spots for quick weekday respites and more convenient alternatives to the municipal forest on weekends (McCreary 1964, 180).

This changing attitude toward city-led greening in Essen coincided with the municipal government stepping into new administrative roles in other areas—all of which was part of establishing the familiar roles and responsibilities of contemporary municipal governance (Lees 2002). From the 1870s, and especially after 1885, when a new mayor took office, the city steadily took over management of activities and amenities from the private sector. For instance, its first response to a cholera epidemic in the late nineteenth century was to make rules about keeping streets clean; early in the twentieth century, it assumed responsibility for cleaning the streets as well as removing trash, maintaining sewers, and cleaning public squares. And, though the city was not yet building public housing, it did implement building inspections and regulations (McCreary 1964, 105, 184). It also began to build cultural institutions such as libraries and opera houses. Taking their lead from the industrialists—and generally with their financial contributions as well—public-sector officials built the Free Public Library in 1902 (a second, Krupp having already provided one) and, between 1906 and 1909, an orchestra hall, a theater, and a municipal hospital (McCreary 1964, 151–53, 174). These efforts have all been understood to be part of "the city's evolution toward a modern 'metropolis,'" along with professionalizing the municipal bureaucracy, turning the old town center into a cultural city center, and founding regional organizations to handle regional-level environmental problems (Von Petz 1990, 4).

By the turn of the century, Essen's parks and green spaces were seen as offering opportunities for leisure. A local 1902 guidebook describes the parks familiarly—and in much the same terms as Margarethenhöhe—as urban public spaces that offered a pleasant respite from the city and opportunities for social mixing. In their "shady gardens . . . the different classes of the population could recover united and happy," away from the "dust and dirt of the city" (Kellen 1902, 146–47). In an advertisement in the guidebook, a local restaurateur promotes the "exquisite" food, company, and service for Essen's "honored residents and the traveling public" in restaurants in the Stadtgarten and the Nordpark (Kellen 1902, 23). As the responsibility for greening shifted from the private to the public sector, nature, newly defined as a public good in the early years of the twentieth century, became a resource for meeting spiritual and intellectual needs, a place for promoting and enacting new city behaviors, and an opportunity for cultural consumption and leisure for all.

## Green Space as Urban Form

Green space was used to enhance urban form as well cosmopolitan culture. Von Petz (1990) has characterized Margarethenhöhe as an accomplishment of unified planning that marked the emergence of new twentieth-century planning principles in the Ruhr. Each tier of the garden suburb's green space, from individual yards to the preserved woodland and meadow that buffered the settlement from the city, was planned in an ordered relationship. This understanding also became a planning charge for the entire Ruhr region, in an effort led by Robert Schmidt under the banner *Grünflächenpolitik*, or "green space policy." Just as positioning Margarethenhöhe as a suburb helped define the city of Essen at the regional scale, Schmidt used large green corridors between the Ruhr's cities to define an urban form that suited the area's polycentrism.

Schmidt still had to advocate strongly for regional green space as a solution to the Ruhr's problems of urban form. As municipal architect he was said to be "decorating" Essen's streets with trees in 1901; in 1906, he became the deputy planner and a member of the board of Margarethenhöhe, whose use of green space he influenced and applauded (Von Petz 1990). When, on the basis of his successes in Essen, he was commissioned to draw up a regional development plan in 1910, he was able to raise the "green space question" (*Grünflächenfrage*) on a larger scale (Von Petz 1999, 166). Unlike most planners of his day, he already saw "park planning as regional planning rather than design"—as engineering rather than decoration—and used green space to create "spatial order" in the sprawling, chaotic Ruhr (Von Petz 1999, 166,

169, 170). He began working to preserve and enhance the distinction between the Ruhr's residential and working quarters through green corridors and arterial roads that defined and created connections between them. His *Denkschrift* emphasized the need to preserve existing green spaces between and within the city centers so that the Ruhr could avoid the problems of urban density that Berlin was facing at the time, arguing that it was easier to preserve open space now than to "free it from buildings later" (Schmidt [1912b] 2009, 90).

Schmidt had an easier time gathering support for expanding roads than he did making the case for greenbelts, however. He attributed the city's reluctance to the large amount of remaining open space in the Ruhr, which made it hard to sound the alarm bell for conservation, and the cost of acquiring the land (Schmidt [1912b] 2009, 64, 68). But the resistance he encountered also suggests that he faced the same problem as local philanthropists: people still needed help seeing nature as a planning tool. Still, Schmidt's green urban vision ultimately won out. His 1912 report laid out the proposal for the still-contested new green corridors, and, by 1920, a Ruhr regional planning association (the Siedlungsverband Ruhrkohlenbezirk, or SVR) was founded, which Schmidt would lead for its first twelve years. The SVR adopted his regional planning approach and green corridors as its foundation, thus preserving a network of green space in the Ruhr that defined its cities early in the twentieth century and still does today (the vertical strips between cities in fig. 3.1 in the next chapter depict Schmidt's greenbelts in the 1960s).

## THE NEW PATERNALISM

A central argument of this book is that, while greening is a form of social engineering, it appears to both greening protagonists and receiving audiences as the provision of public goods. And, indeed, the new conception of nature visible at Margarethenhöhe and throughout the Ruhr had affinities with a new form of managerial control—in and outside the workplace—that appeared both not controlling and actively benevolent. In America by the 1870s, the feudal paternalism of early factory towns was being replaced by a "new paternalism" motivated by notions of "laissez-faire individualism—self-interest, economic rationality, self reliance" (Crawford 1995, 34). The feudal paternalism was that of the colonies, where industrial barons managed employees understood to lack self-control with carrots and sticks, and fell back on explicit forms of punishment and domination as necessary. The new paternalism was that of Margarethenhöhe, which governed through the appearance of autonomy and self-discipline. It resembled what the sociologist Patrick Joyce has called the "rule of freedom," in describing a form of governmentality characteristic of modern liberalism in which the absence of restraint becomes a "technique of rule" and a "formula for exercising power" (Joyce 2003, 1). For Joyce, urban public life was a key site for the deployment of these technologies. In the Ruhr in the early twentieth century, both industrial barons and city governments came to cultivate liberal subjects through physical and programmatic interventions—schools, libraries, hospitals, health insurance, and cultural and recreational amenities—with the hope that these would both facilitate better health and well-being and greater self-management.

Such strategies were visible far beyond green space. Alongside growing public investment in public spaces and public culture, Krupp began to offer workers new forms of social support in this era. In addition to housing, health care, and retirement, the company began to sponsor education and enrichment programs ("*Belehrung, Fortbildung, und Unterhaltung*," or "instruction, continuing education, and conversation") and provide public education in the city and the settlements (Krupp'sche Gussstahlfabrik 1912a, 205). In 1899, it established a company library and an educational association, the Krupp'sche Bildungsverein—translated in company materials in 1912 as the "Krupp Association for the Promotion of Mental and Social Culture"—which offered classes and cultural events for all employees and the general public (Beitz 1994, 137–54, 160–80; Krupp'sche Gussstahlfabrik 1912b, 309). In 1911, when the *Bildungsverein* held an exhibition about home furnishings that celebrated the lifestyle of the modern worker (Beitz 1994, 197–99), its images reflected the ideal that had been built into Margarethenhöhe: homes with no laundry in the yards, tidy gardens, and warm, wallpapered rooms with photographs on the walls and contemporary wooden furniture.

Nature's new status as a civic, leisure space and its appearance as presocial, uninterested, and unmediated helped make greening an important part of this suite of managerial technologies. Joyce describes nineteenth-century urban infrastructure as helping people have an experience of ecological and moral freedom: helping bodies, water, and air circulate freely so that citizens could have the experience of "natural" physical and "moral self-regulation" without "undue interference" (Joyce 2003, 70). In contrast to the colonies, where nature's power lay in its ability to satisfy material needs, green space as urban infrastructure in Margarethenhöhe and the Ruhr created just this sort of experience. Once the paradigmatic form of engagement with nature became leisure rather than labor, spaces such as parks and barn-cafés became detached from work and oriented toward the sphere of social reproduction. This helped make the idea of nature's universal benefit concrete by making these places for moral and spiritual improvement, demarcated, spatially and temporally, from paid work, and therefore disconnected from questions of class and power. By the beginning of the twentieth century, public and private green spaces in and outside cities (the countryside for Jurgis, the garden for Margarethenhöhe's male breadwinner) provided the experience of a free society, just as did water and sanitation infrastructure or access to museums and libraries.

In other words, with the arrival of urbanized nature, greening's managerial mode changed—in Gramsci's terms, from coercive forms of domination to hegemony and the manufacture of consent. Greening created apparently free and unmediated spaces that were nevertheless sites for learning behaviors that upheld the status quo and that materially instantiated a social and political regime through the rhythms of daily life. Rather than kitchen gardens that prevented workers from "throwing a glance over the garden fence into the world" by chaining them to the worries of everyday provisioning (Treiber and Steinert 1980 quoted in Steinborn 1991, 12), by the twentieth century gardens as leisure spaces became "one of the best means to elevate the mind above the worries of material existence" (Ladd 1990, 69). Private gardens offered "a simultaneous domestication of working men's leisure and of working women's productivity" (Van Zee 2012, 2), while public parks were places for learning bourgeois culture, mingling with the new middle class, and putting oneself on display.

While the provision of nature in the colonies was also part of providing satisfying and fulfilling lives for workers, the transformation of nature into a civic, leisure space strengthened nature providers' understandings of greening as acting in the public good. For the private sector, it helped turn greening into a philanthropic practice. For the public sector, it made greening part of the package of urban amenities—along with basic water and sanitation infrastructure—that municipal governments increasingly expected to provide. The fact that Margarethenhöhe was conceived as a gift to the city illustrates this point. The Ruhr was an important site of industrial strikes and labor organizing (Crew 1979; Nolan 2003), and Krupp was notoriously heavy-handed in managing employee unrest. Company housing was transparently part of his disciplinary tool kit. But Margarethenhöhe was articulated as the provision of a public good, a gift not just to Krupp workers but to all city residents.

The latent managerial aspects of these green spaces do not mean that target audiences would have been displeased by their new access to nature. Even if they were a technique of control, Krupp, Osthaus, and Schmidt's efforts to make green space accessible to industrial workers democratized leisure experiences of nature, something that the working and middle classes were likely not to have found all bad. And, while it is not clear from the historical record whether members of the public received green gifts as they were intended, another mark of the installation of this new common sense is that greening protagonists at least acted as though they would. The Ruhr did share one characteristic with traditional cities: a huge percentage of its workforce was foreign born. But providers of nature did not express concern that anyone—from unskilled workers to white-collar managers, from a variety of backgrounds across Europe—would have trouble understanding Margarethenhöhe as a philanthropic act or its gardens as especially desirable and beneficial.

### GREENING'S MANAGERIAL UTILITY

As Krupp, Osthaus, and Schmidt gave their new urban vision of society and understanding of nature material form at Margarethenhöhe, they were greening in the familiar, contemporary sense, and their actions were predicated on the new availability of urbanized nature rather than any local conditions. It is historically and theoretically relevant that protagonists were able to green the Ruhr even in the context of low-density polycentrism because, for all intents and purposes, the region's colonies already provided just what the garden city offered: small clusters of multifamily units joined by arterial roads and buffered by green space. But early twentieth-century actors did not green the region to alleviate the density and public health problems that plagued more conventional industrial metropolises; they greened because they wanted to make the Ruhr into a city. They were able to do so because they treated greening as a social practice that contributed to the organization of urban life and could be replicated even in an area where much of the housing already offered low-density units with ample green space. They also assumed that greening projects would not only help make the Ruhr appear urbane to international audiences but also be beneficial and transformative for people not living traditionally urban lives—that is, help turn Krupp's employees from people perceived to be like peasants into cultured cosmopolitans.

This new understanding of nature, new normative vision of society, and new set of social practices were inaugurated at Margarethenhöhe. In contrast to the colonies that preceded them, garden cities in the Ruhr reflected a distinctly *urbanized* imaginary of nature in that they used nature as a new kind of indirect, aspirational, universal good. To be clear, both versions of nature are social products. The contrast between the colonies and the garden cities is not between a first and a second nature, between an ontologically real nature and a social artifact. Rather, garden cities illustrate a collective reorientation toward quotidian forms of nature *as* social products. In the garden city, the same materials came to be used in qualitatively new ways and were filled with new kinds of social content. From being preservationist in spirit and fulfilling direct, material needs, nature's use changed, becoming aspirational and symbolic: green became a medium, a bearer of indirect social goods.

Urbanized nature also gave domestic forms of nature a new kind of managerial utility—as a civilizing force that imposes norms of behavior and citizenship as it transforms cities as physical and social spaces. It is no great surprise that the garden city should have been a site of bourgeois subject formation, given the popularity of urban gardens and green spaces among nineteenth-century urban social reformers. What is interesting is that these methods could be adopted even in the low-density Ruhr. That Krupp and his contemporaries could green even in such an environment also reveals that urbanized nature was not dependent on the classic physical form of the city for experiential contrast—green against gray—to be legible as a vehicle of new social ideals. The design shift from faux-rural colonies to protourban garden suburbs reflected a new normative vision and way of perceiving and using green space. And this change was desirable and the new social contents intelligible even though the two architectural and landscape forms were almost identical; garden cities and greenbelts could be understood to bear new social ideals in spite of their formal similarity to the colonies or even Fritsch's garden cities.

Arguments about the "modularity" of social imaginaries have emphasized how imaginaries transform as they travel in terms of substantive differences in content (Goswami 2002). Anderson (2006), for example, distinguished between Creole, linguistic, and official nationalisms adopted in different contexts. To this we might add that social imaginaries are modular in terms of use. As they travel to new places with different local conditions, they are put to work in new ways and find new social functions. As industrial barons looked to Paris and Berlin as models around 1900, they saw the garden city as a design that had potential use in the Ruhr's landscape in spite of the fact that the region did not lack green space. Instead, they used the same design and conception of nature as a solution to very different problems: to make the Ruhr into a city.

In tracing urbanized nature's historical emergence, this pair of chapters underscores the extent to which a number of things we are prone to naturalize today as essential characteristics of nature—its moral benefit, its universal appeal, its status as a leisure space—are not timeless but quintessentially modern, urban, and historically specific understandings. At the same time, Margarethenhöhe also serves to establish these characteristics as basic features of greening as a social practice more generally. They appeared in the Ruhr at the beginning of the twentieth century, but they are inherent to this imaginary and to greening practices as they persist today. One of these is greening's combination of paternalism and good intentions. Margarethenhöhe was a top-down, disciplinary project as much as a gift to the city and to Krupp workers, and new associations with nature as a universally beneficial public good and leisure space helped sustain these dual purposes. As the following chapters will show, the beliefs surrounding nature that emerged in the Ruhr at the turn of the twentieth century continued to contribute to greening's managerial utility and shape the dynamics of greening projects' production and reception in the following decades, across changing actors and social relationships.

# * 2 *

# *Contested Social Ideals*

CHAPTER 3

# The Space-Time of Democracy: Parks as a Bourgeois Public Sphere

Once available, urbanized nature and the new uses of nature established at Margarethenhöhe were used to spatialize changing ideals of urban publicness in subsequent eras. A second major greening moment in the Ruhr—and a time when questions of urban public life were again top of mind—occurred fifty years later, in the decades after World War II. Though the 1960s and 1970s were a time of national crisis in West Germany—of recovery from the horror, shame, divided country, and total devastation left in the wake of National Socialism (during which representations of nature had also served other, nationalistic purposes)—they were also a time of urban crisis, for several reasons. First, National Socialism was understood to have brought about the "irreversible death" of urban public life (Moses 1999; Prowe 2001; Salin 1960, 23–24), leaving urban practitioners, social theorists, and members of the public preoccupied with the question of how to rebuild democracy in postwar West Germany and the opportunities afforded by urban public space for that project. Second, there were worries about the growing amount of free time and new forms of public leisure and consumption. By the 1960s, the new consumer goods and domestic efficiencies of West Germany's postwar "economic miracle" (*Wirtschaftswunder*) had improved daily life, but they had also sparked concerns about how that free time should best be spent. Finally, in the 1970s, with the global decline of Fordism, the industrial Ruhr began to experience the collapse of its coal and steel economy. The region was almost completely reliant on these industries, and concerned local and regional governments responded by beginning to plan for a different economic future.

In this context, the Ruhr again turned to green space. As throughout West Germany, greening projects became key sites for playing out solutions and imagining alternative social and political futures in the landscape, through interventions that targeted public spaces and the people who used them. Nationally, there were two competing political visions for postwar democratic life: a more mainstream, reformist ideal of the 1945 generation that had experienced National Socialism, and a more revolutionary one associated with the 1968 generation and the New Left. In the Ruhr in the 1970s, two new groups of protagonists mobilized urbanized nature to advance greening projects embodying these different visions of urban public life. One was the "skeptical" '45ers (Mitscherlich and Mitscherlich 1975; Moses 1999), who aimed to "politicize the public sphere" in "a new climate of polite pluralism, national civility, and reasoned public exchange" (Eley 2002, 227) and promoted a series of regional recreation parks, called *Revierparks*, that projected a Habermasian vision of a bourgeois public sphere. The other was the '68ers, who promoted a proletarian, New Left vision that they found in the Ruhr's traditional workers' colonies and that drew on the work of two of Habermas's students, Oskar Negt and Alexander Kluge.

The coexistence of these two very different political visions—both expressed through green space in the same time and location—makes clear that urbanized nature is a condition for action rather than any specific normative idea of society. The projects' divergent critiques of the present, imagined futures, and relationships to existing structures of power also highlight the radically different social roles that can be assigned to nature in urban politics even in the same period.

Chapter 4 will return to this point. This chapter focuses on the *Revierparks*, examining how social scientists of the day and the planners responsible for the parks' design and construction understood them to represent a Habermasian model of urban public life and perceived their social functions and benefits. Regional planning documents and materials produced to introduce the parks to users show that urbanized nature once again allowed a new set of protagonists to improve the Ruhr by greening it and that nature was still viewed as an indirect, aspirational, and universal good that worked through the rule of freedom. This chapter also introduces a new dynamic regarding nature's perceived universality. The concerns of the 1960s about rebuilding urban democracy highlight how the spatiotemporal and phenomenological qualities of green space, as perceived through this social imaginary, made it easy to experience green spaces both as socially neutral—by virtue of their separation from work and home—and as physically pleasant and desirable.

These qualities made green spaces especially useful for creating democratic urban publics in the Ruhr in the 1960s and, more generally, help explain why green space is a favorite site for democracy in cities.

In the late 1960s and the early 1970s, the Siedlungsverband Ruhrkohlenbezirk (SVR), the regional planning organization founded by Robert Schmidt in 1920, planned and built a series of five large regional recreation parks in the Ruhr. The eighty-acre parks boasted large grassy areas, playing fields, and year-round facilities for camping, swimming, horseback riding, ice skating, and weight lifting. People could visit a sauna or a pub, participate in music and arts programs, or simply engage in "quiet relaxation" (*Erholung*) in the grass and sunshine (Landesregierung Nordrhein-Westfalen 1968, 65). The parks—which still exist and, at the time of this writing, are being updated as valuable green infrastructure serving ecological as well as social functions—were, in the 1960s, a new kind of park, built in direct response to the political concerns of the day. The *Revierparks* continued the tradition of *Volksparks*, or "people's parks," built in Germany beginning in the 1920s, in that they provided convenient recreation opportunities for the working class. But, unlike prewar parks, which were also "an expression of the spiritual unity and cultural identity of the German nation" (Tate 2013, 105; see also Gaida and Grothe 1997, 29), after World War II Ruhr planners reconceived *Revierparks*' users and politics, understanding them as a new kind of *Volkspark* for a new kind of inclusive, democratic *Volk*.

A 1972 book by the German sociologist Hartmut Lüdtke called *Freizeit in der Industriegesellschaft* (Leisure time in industrial society) described the Ruhr's *Revierparks* as a new national prototype of beneficial leisure space—one that promoted "an open and democratic society" and the "'consumption' of landscape, green, light, air, and sun" rather than material goods (Lüdtke 1972, 39, 49). This chapter examines social science literature of this kind, as well as planning and promotional documents for the new parks, in order to understand what purposes planners understood the parks to serve, what benefits they understood them to be providing, and how these projects fit in with larger social, political, and economic goals. These documents reveal that, in the context of a local and national conversation about urban democracy, leisure time, and the identity of the Ruhr and its people, *Revierparks* were explicitly and literally understood to be settings for democratic public life. Though not unique to the postwar era, West Germany's political concerns in the 1960s brought those dynamics especially close to the surface. Rather than expressions of German cultural identity or "escapes" from society, the parks were treated as new space-times of democracy: places in which to create a revived public realm, to experience diversity and heterogeneity, and to enact new bourgeois subjects and forms of public politics.

## PARKS AS A PUBLIC SPHERE

A key influence on the *Revierparks* was Jürgen Habermas's *The Structural Transformation of the Public Sphere*, which was first published in German in 1962. At the time of its publication, the book's historical analysis was motivated by contemporary questions about the postwar public realm and part of a live "debate about desirable meanings of publicness—*Öffentlichkeit*—for West German democracy" (Eley 2002, 227; see also McCarthy 1991, xii). Habermas's now-classic account of the rise and fall of the bourgeois public sphere also provided a model for postwar politics (Calhoun 1992; Eley 1992; Kramer 1992). In the 1960s and 1970s, German planners and policy makers tasked with rebuilding democratic public life read and were influenced by Habermas's work, along with that of now lesser-known social scientists analyzing problems of the day—such as the rise of leisure time—and linking them to the then-dominant modernist planning paradigm. Like Osthaus's essays on the garden city fifty years earlier, these scholarly works were widely read and influenced public debate as well as urban policy and practice. In addition, they articulated the social and political ideals spatialized through greening projects. As such, they are used as primary sources in this chapter and the next.

As a work of social theory rather than a historical source, *The Structural Transformation of the Public Sphere* is also useful for helping clarify how and why problems of urban democracy were addressed through green space—why parks might have been the answer to publicity as an "organizational principle" in postwar West Germany (Habermas 1991a, 4). For the public sphere is not always synonymous with public space and need not be urban. It can be an abstract social space or media-based sphere (Warner 2002); for Habermas, it was less a specific vision of concrete places than an abstract framework for democracy, a set of procedural norms, a way of making collective decisions, an agreement about the exercise of power (Calhoun 2002). But place helps create publics; places—especially public places, and even more so urban public places by virtue of the density and heterogeneity of those who inhabit them—present opportunities for publics to form and settings in which procedural norms are played out. And, as Lüdtke's characterization of the *Revierparks* as a "prototype" of spaces that could contribute to "an open and democratic society" suggests, urban public space has been central to the actual formation and transformation of the public sphere and public life beyond modern urbanism, from the Greek agora to more contemporary churches, marketplaces, and village squares (Mitchell 1995; Young 1990, 226–56). West German planners using Habermas as a model for thinking about urban public space came to understand public green spaces as sites for addressing abstract concerns about the public sphere and places to actually create one.

Habermas's influence on planners also illuminates the extent to which greening is, more generally, a strategy of spatializing urban publics and how greening projects have frequently done so by reconfiguring the relationship between public and private spheres. Urbanists in the interactionist tradition, in particular, have described urban history as successive reorganizations of public and private, by examining the design of new interactional environments for new social orders (Goffman [1963] 2008; Jacobs 1961; Lofland 1998; Sennett 1992; Whyte 1980). Take Richard Sennett's *The Fall of Public Man* (1977), which traces the role of urban public spaces in the constitution of a public realm in Europe. Sennett begins with the ancien régime, the period when "commercial and administrative bureaucracy grew up in nations side by side with the persistence of feudal privileges" and the urban bourgeoisie were still perceived as rulers, administrators, and sources of financial support. He documents a growing focus on public life through the eighteenth century as the "cosmopolis" was created in European capital cities through investment in large public spaces where "even the laboring classes [could begin] to adopt some of the habits of sociability" for the first time. These forms of public life were "inherited culture" for nineteenth-century civic leaders, who copied it even as that civic world was "crumbling," by providing a new urban bourgeoisie and eventually all city residents access to the parks, pedestrian streets, coffee houses, theaters, and opera houses that were "formerly the exclusive province of the elite." The "fall of public man" occurred as the public realm came to be understood as a potentially immoral, dangerous domain in contrast to the home as a private refuge. Sennett describes the twentieth century as ushering in "the end of public culture," as familiarity and tight social ties came to be prized over of public interactions and urban public space became something to fear (Sennett 1977, 47, 18, 141, 17, 259).

Similarly, greening interventions produce new urban geographies by reconfiguring the relationship between public and private and defining new sets of public behaviors in each era. Sennett's time line harmonizes with Habermas's account of the rise of the bourgeois public sphere and traces the historical transformations covered in part 1 in a green register. The ancien régime was the fiction created by the world of colonies, reflected in the Ruhr's industrial barons' insular housing design, lack of interest in public space, and "feudal" management style, all of which provided an anachronistic representation of social relationships through the end of nineteenth century. In the beginning of the twentieth century, the garden city created the first truly public life in the Ruhr, modeled on the European cosmopolis, and characterized by a new division between public and private and public appearances organized around the consumption of goods and culture. The new understanding of nature as a public good made it possible for planners and philanthropists to create a public sphere through green space as well as through investments in libraries, museums, and lecture halls. In the 1960s and 1970s, Ruhr planners turned again to green space as a site for constituting an urban public, though in a form that articulated differently with other realms of life. They preserved the home as a private refuge and saw parks as places for private people to come together as a public.

This chapter's focus on green public spaces also suggests that evocations of the public good in the context of green space are especially powerful because of nature's imaginative and material power. Nature's construction as a universal good conditions a view of urban parks and green spaces as morally and socially beneficial in spite of their shortcomings. Actually existing urban green spaces will, of course, be less universally public than they aspire to be, reflecting the material constraints, forms of power and rationality, and normative visions of their era as well as the blind spots and biases of their creators. Yet greening projects are nevertheless marked by their creators' perpetual public *ambitions* and are perhaps unusual in the degree to which these ambitions can be sustained. In comparison to other public goods—such as knowledge, security, and culture, provided through institutions such as schools, police, and museums—it is perhaps easier to experience green space as presocial and transparently beneficial, both because urbanized nature so strongly primes users to perceive greening projects in these terms and because the fact of green spaces' social construction does not always affect the experience of their use. An ornamental tree planted on a lawn casts no less shade than a native species stumbled upon in a grassy meadow; both provide equal relief on a hot summer day even if one is more obviously a planned, planted human construction.

## THE POSTWAR CRISIS AS AN URBAN CRISIS

Urban planners moved to spatialize a new public sphere through *Revierparks* in the 1970s because they were responding to a set of problems being articulated by social scientists as urban problems and to which spatial planning had been posited as a solution. Life in the Ruhr was grim in the immediate aftermath of World War II. Local industry and transportation networks had been heavily bombed because of the region's central role in arms production. According to one lifetime resident I interviewed, the destruction was so great that during this era even Margarethenhöhe had to lift its restrictions on gardening so its residents could grow food. But, by 1950, West Germany's economic miracle had begun, thanks in no small part to Ruhr coal. In 1951, for example, the Ruhr's state of North Rhine–Westphalia (NRW) was responsible for 40 percent of the Federal Republic's postwar industrial production and 90 percent of its iron and steel (Goch 2002a; Roseman 1992). Nationally, productivity was on the rise, real incomes were growing, there were massive investments in infrastructure and public services, and the population was growing steadily. In Germany, as throughout Western Europe and North America, the postwar boom was an era of mass production, suburban expansion, family cars, and affordable domestic and consumer goods, all of which were creating a new landscape of middle-class wage earners and of mass consumption (Cohen 2003; Jessop 1994).

By the mid-1960s, West German social scientists began expressing concern about the social effects of the increasing wealth, consumption, and spare time of this postwar "leisure society." Household work was being increasingly outsourced and standardized, labor legislation shortened working hours, Taylorist management and the automation of production boosted productivity, democratization prioritized leisure forms of play and domestic life over work as the "morally binding center of worth of life," and new forms of mass marketing and consumption became ubiquitous (Lüdtke 1972, 14). As a result of these changes, leisure time enlarged to become, for the first time, a defined temporal realm separate from both work and home (Herlyn 1970, 23–26). In West Germany the new "phenomenon" of leisure—evenings, weekends, and holidays—became a research object (Huck 1980) and a social problem (Grothe 2003; Schildt and Schildt 1996; Weber 1963). "Everywhere," "doubts were raised as to 'whether mankind was really prepared for so much free time'" (Weber 1957 quoted in Schildt and Schildt 1996, 220), and there was a growing concern about how it would be spent—and how "cheaply," in the ever-expanding realm of consumption and advertising, "freedom and leisure time could be sold" (Hezel 1974, 138). If properly managed, *Freizeit* (leisure time) could be a space for human freedom and physical and spiritual rejuvenation. If not, the availability of time and money risked degeneration into vacuous *freie Zeit* (free time) that was aimless, unmoored, or solely consumption oriented (Lüdtke 1972).

These concerns arose during a moment of Fordist modernization that also had a distinct spatial character. German planners engaged in postwar reconstruction embraced architectural modernism and functionalist planning—a stark departure from the ideal of the centralized nineteenth-century industrial city. Internationally, this planning paradigm was best known architecturally through Le Corbusier's vision of "skyscrapers in a park," programmatically through the 1933 Athens Charter, which laid out a vision of functionally divided cities, and practically through the construction of Brasília, the new capital of Brazil, in 1960. In West Germany, high-rise apartment blocks on cities' outskirts were initially an efficient and affordable solution to the postwar housing crisis (Diefendorf 1990). But, after about 1965, planners embraced the modernist city as they turned to the future. As Christopher Klemek has drily put it: "Having already rebuilt areas destroyed during the war in keeping with modernist principles, German urbanists turned in earnest toward eradicating the surviving remnants of the nineteenth-century cityscape" (2007, 56). Legislation assisted in this. A 1960 Federal Building Act (Bundesbaugesetz) supported a shift toward centrally administered, regionally scaled urban planning by extending municipal authority in city planning and redevelopment. It was followed in 1971 by a Städtebauförderungsgesetz—variously translated as the Urban Improvement Act or the Urban Development Assistance Act—which dedicated formal renewal areas in old portions of the central cities and increased state and federal funding for urban redevelopment programs (Friedrichs 1987; Goch 2002b, 294; von Einem 1982).

German social scientists described the negative effects of modernist planning and mass consumption as the decline of "urbanity," and tapped urban planners for solutions. In the 1960s, books such as Hans-Paul Bahrdt's *Die moderne Großstadt* (The modern metropolis) ([1961] 2006) and Alexander Mitscherlich's *Die Unwirtlichkeit unserer Städte* (The inhospitality of our cities) (1965) began to raise questions about the social and psychological costs of large-scale, anonymous residential and public spaces, which scholars saw as threats to the fragile postwar public realm. Bahrdt and Mitscherlich both worried about "streets and squares [that had] lost their public character . . . since they have come under the dictates of total traffic" (Bahrdt 1952, 1476). The economist Edgar Salin (who also happened to be one of Talcott Parson's dissertation advisers) wondered how to revive "urbanity" in an environment of mass culture (Salin 1960). Whether the comparative referents were the nineteenth-century European cosmopolis or a vision of small-scale "community" that "rested on social-psychological rather than national-cultural criteria" (Koshar 1998, 227; see also Mitscherlich 1965), social scientists saw leisure time and Fordist modernization as threats to public life. They linked concerns about the public sphere to broader questions about the spatial and temporal organization of urban life and used spatial planning as well as procedural interventions to help address them. Legislation helped secure German planners' commitment to "humane urban development" (Bahrdt 1968), for instance, by mandating that urban renewal be accompanied by "social plans" and carried out with citizen participation (Friedrichs 1987; von Einem 1982).

Concerns regarding the relationship between the spatial organization of the urban environment and its social and political character produced a close relationship between sociology and urban planning in Germany in the 1960s (Herlyn 2006, 7). This was a collaboration driven above all by efforts to "confront Habermas's [public sphere] thesis" and address concerns about the public sphere's "decay" by planning cities that were hospitable to the formation and maintenance of a genuinely democratic society (Bahrdt [1961] 2006, 30). As a result, sociology—and the data sociologists gathered—came to be understood as "the servant of good planning" in West Germany (Klemek 2011, 125), and, by the 1970s, social scientists were arguing that "the public sector [had] to take on significant work in [the area of leisure time]" (Hezel 1974, 137). Their recommendations, as outlined in a 1974 text, included providing more and more diverse urban leisure-time opportunities; seeing the provision and financing of such activities as functions of urban government; securing the open space necessary for this "future leisure society"; "opening up" green and open space for leisure-time use; establishing leisure activities as opportunities for communication; providing better access to these new resources via public and private transportation; and ensuring their equal distribution and equal access—all with the goal of "creating a citizenry" (Hezel 1974, 137).

Locally, the Ruhr was also confronting specific problems related to the collapse of its industrial economy. In spite of Schmidt's, Osthaus's, and Krupp's efforts early in the century, the region had not become the hoped-for international model of prosperous middle-class urbanism. Though, as in 1900, the Ruhr's unusual morphology gave it some unique strengths that might have helped overcome its negative public perception—its decentralized character (Hall 1966) and proximity to Bonn, the new political center and administrative capital, just one hundred kilometers away—in 1960 coal production sent profits elsewhere while blackening local skies, and contemporary accounts still found the region hopelessly chaotic, polluted, and provincial. Though daily life improved with the *Wirtschaftswunder*, the region was hit hard by a coal crisis in 1959 as oil and gas in the Middle East and North Africa began to create an affordable alternative to coal. This marked the beginning of both the end of the coal and steel economy and a process of deindustrialization that would continue into the twenty-first century. Even as the *New York Times* reported in 1965 that the roads in the Ruhr were "jammed with automobiles" and "the houses . . . packed with television sets and washing machines" (Shabecoff 1965), employment dropped from a peak of 468,000 in 1957 to 215,000 in 1968 (Spelt 1969, 4), and between 1966 and 1967, 113,388 people emigrated out of the Ruhr (Benedict 2000, 71). As the local economy began to collapse, a 1968 report on the Ruhr and its coal industry announced:

> A most serious handicap in attracting new industries is that the Ruhr has an unfavourable image. In the minds of the public, the district is identified with coal mining, heavy iron and steel, and a landscape to match. This is a more serious problem for its continued growth than the coal crisis. The Ruhr lacks many amenities compared with other industrial locations in Germany or across the border. Coal mining has moulded and shaped the Ruhr landscape, which in many respects is harsh and ugly. It is a scene characterized by mine heaps, air pollution, destruction of vegetation, land subsidence, lowering of the groundwater table, water pollution, dreary workers' settlements, and so forth. Towns like Bottrop or Wanne-Eickel are not really cities. In order to attract and retain migrant workers from distant places, the mines built housing developments on their own land, often at great densities. Public services in these towns are of a bare minimum. They do not have any of the amenities commonly associated with urban living. It should be stressed however that the Ruhr does not contain any real slums. Its worst compares favourably with what one sees in the mining areas of England or France. Yet with the closing of the mines these towns must attract industry or become dormitory settlements. They are suited for neither. (Spelt 1969, 8)

Similar perceptions of the Ruhr's local residents also persisted. The weekly news magazine *Der Spiegel* ran an evocative cover story on the "everyday misery" of life in the Ruhr in 1961, the year of the region's peak population and peak pollution ("Zu blauen Himmeln" 1961; see also Brüggemeier 1994; and SVR 1969, A13). Its impetus was the Social Democratic Party's candidate for chancellor, Willy Brandt's, campaign promise to "make the skies over the Ruhr blue again" by regulating industrial emissions and waste disposal. Most of the article parodied these ambitions with a detailed account of the Ruhr's wretched environment. It described a "plague" of airborne dust and ash ten times higher than that in Berlin that reduced the strength of the sun's rays by one-third, caused over one hundred deaths from lung cancer each year, forced waiters in fine restaurants to change their shirts three times a day, and prevented housewives from hanging laundry outside to dry except during favorable wind conditions. It also depicted the Ruhr's people: "The sky in the Ruhr was now always 'sold out' [of sun], and when the miners and steelworkers gathered in noisy flocks on the Rhine and Mosel during the summer on the weekends, they tried to make up for the year's lost sunshine after the fact—and on tap. Occasionally, and especially after returning from such lovely holiday landscapes, they, and especially their wives, were overcome with the so-called 'Ruhr rage'" ("Zu blauen Himmeln" 1961, 24–25).

As in the first decade of the twentieth century, such descriptions raised two concerns for local planners. Planning documents of the era reflect a desire to change the Ruhr's reputation in order to help attract new businesses, such as the NRW's 1970 regional plan, which describes a "beleaguered industrial landscape of crisis" transforming into one of Europe's "hopeful economic landscapes of the future" (Landesregierung Nordrhein-Westfalen 1970, i). But, obviously, deindustrialization affected individuals differently, and, in the traditionally working-class Ruhr, it was the coal miners and steelworkers and their families—the majority of the population—who would disproportionately feel the effects. And so there was also the question of what to do with these people, and how not to leave them behind. As at Margarethenhöhe—though now for material rather than ideological reasons—there was a need to turn an uncultured working class, an unfavorable public image, an "ugly" and polluted industrial landscape, an absence of "real cities," and a lack of urban amenities into a place and a people with a postindustry future.

## SPATIALIZING THE URBAN PUBLIC SPHERE

It was in this context of national economic prosperity combined with local economic and environmental precarity that the Ruhr's public sector adopted a "modernization and future-oriented" (Goch 2002b, 294) development program aimed at rebuilding a public realm while transitioning to a postindustrial, leisure-oriented economy. As the industrial barons' local power waned with the collapsing industrial economy, the public sector stepped in, taking up the '45 generation's Habermasian vision, "however flawed" and "idealized" (Eley 2002, 226), and making the creation of a "pluralistic leisure society" its central task (Grothe 2003, 39).

The public sector was actually quite well positioned to begin acting on these goals, and developed ambitious regional plans to address them. The SVR was still dedicated to comprehensive regional spatial planning, and the NRW was in the process of developing one of the most ambitious planning programs of all the West German states (von Einem 1982, 19), one that prioritized investment in the struggling Ruhr over areas such as Düsseldorf and Bielefeld (Landesregierung Nordrhein-Westfalen 1968, 55).* These regional planning bodies, supported by federal legislation and funds, led local planning efforts. In 1966, the SVR published a regional development plan outlining thematic and geographic priorities. In 1968, the NRW published a comprehensive five-year Ruhr development program, which was followed in 1970 by a plan for all of North Rhine–Westphalia that carried many of these ideas forward. In such documents, planners took responsibility for creating a public realm for democracy while addressing structural change, the Ruhr's bad reputation, and pressing environmental problems.

Barred from turning to history for "national-historical identity formation" (Habermas 1988, 5), planners looked to American cities for solutions. The United States provided inspiration for a new West German economy, new political institutions, and new forms of urban space (Almond and Verba [1963] 2015; Habermas 1988, 1991b), and its automobile-oriented cities appeared to be the spatial form synonymous with Western democracy. Thus, in spite of the worries about modernist planning's effects on urbanity, and while critical of the autocratic style that often accompanied large-scale planning in the United States, most notoriously Robert Moses's work in New York (Diefendorf 1999; Klemek 2011), Ruhr planners adopted functionalist planning principles. This marked their participation in a transatlantic postwar urban renewal "order" characterized by modernism in architecture and functionalism in planning, the professionalization of planning as a discipline, increased public-sector involvement at the regional and federal levels, and ambitious, large-scale urban redevelopment schemes (Klemek 2011, 19).†

---

* Locally, the pressures of structural change had also produced an unusually strong alliance between local and regional government, the private sector, nongovernment organizations, and the public as all rallied together to cushion the blows of the collapsing coal and steel economy and prevent local economic and social catastrophe (Goch 2002a; Kift 2013). The SVR also had the opportunity to have its decisions guided by a number of social scientific studies of its citizens carried out at the Dortmund Social Research Center (Sozialforschungsstelle Dortmund) at the nearby University of Münster in the 1950s as the Ruhr's blue-collar workforce attracted researchers interested in questions of class and political consciousness and changes in workers' conceptions of society (*Gesellschaftsbild*) as a consequence of rationalization, technological advancement, and structural change (Brepohl 1957; Jantke 1953; Popitz, Bahrdt, Jüres, and Kesting 1957; Schmidt 1969).

† As in 1900, these solutions were adopted in the Ruhr in the absence of the specific problems motivating their use elsewhere. The march of functionally divided cities across the United States was a response to white flight, urban blight, and black poverty and was marked by a strong

Spatially, planners' vision of democracy looked like a functionally divided city. The planning establishment responded to Habermas's democratic vision and concerns about modernist planning and rampant consumption by turning urbanity into a normative and prescriptive category for social life (Diefendorf 1990, 279; Herlyn 2006, 7, 16). West German planners "invoked" urban ideals by using terms such as *shared public space*, *the public realm*, and *public grounds* (Klemek 2011, 91). They defined urbanity as "a cultural quality" that "develops out of the ethnic, social, and political heterogeneity of the population, out of the tensions between different class situations, life forms, interest orientations, wishes, appetites, and also out of productive discontent and calm tolerance" in public space (Diefendorf 1993, 364). And they operationalized it in a manner compatible with modernist planning, basing it in a functional separation of spaces in which different age, class, and cultural groups could come together. Unlike the dense, walkable downtowns and mixed-use neighborhoods of the cosmopolis, the international "recipe" that was "evangelized as the sole possibility for enlightened city building" (Klemek 2011, 29) in this era was an "articulated and 'loosened' city" (Göderitz, Rainer, and Hoffman 1957) that contained four "functionally articulated" zones, for work, transit, private life, and leisure (see Diefendorf 1999, 7–8), as outlined in the Athens Charter. Open space planning was directed at keeping these functional areas "separate and at a distance from each other" (Kastorff-Viehmann 1998, 129). In this new, functionally divided context, urbanity was understood not just as strangers mingling but as a particular "tension" and dynamic created by the separation between the public and the private spheres (Bahrdt [1961] 2006, 86).

As in 1910, green space was central to making this urban future visible and concrete. At the regional scale, greening was still used as a spatial ordering tool. By 1966, the Ruhr was the "biggest contiguous industrial region in West

---

division between top-down planning programs and a bottom-up community backlash against totalizing, large-scale urban redevelopment schemes (the battle between Robert Moses and Jane Jacobs over the proposed Lower Manhattan Expressway being the most famous of these antagonisms). In West Germany, a progressive, reformist alliance of planners, social scientists, and regional policy makers led postwar planning with a national commitment to participatory democracy and public life that kept German planners in dialogue with the public and with social scientists and made professional and community interests relatively less polarized (Klemek 2011). As the Moses-like euphoria about large-scale planning's power to redesign the social and physical landscape waned (Albers 2006), citizen protest—such as those by squatters' movements in Berlin—and the commitment to democratic civil society paved the way for a gentler program of "cautious" urban renewal in the late 1960s and 1970s (Holm and Kuhn 2011) that was more sensitive to historical architecture and contemporary social context.

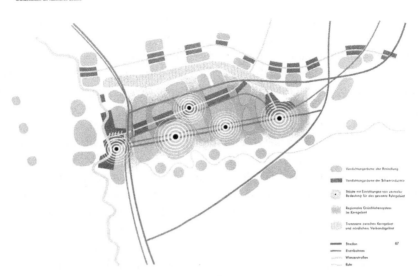

FIGURE 3.1. The Ruhr's functional division of spaces. (From SVR 1966, 67.)

Germany," housing 5.7 million people (10 percent of the country's overall population) on 2 percent of the landmass (SVR 1966, 67, 19). It had a population density of 1,243 adults per square kilometer, as opposed to an average of 225 in the country overall (SVR 1966, 19). The SVR's 1966 regional development plan established "spatial and substantive segments" for different areas (SVR 1966, 2), while the NRW's 1968 plan concentrated housing, facilities, trade, and culture within each of the region's cities and improved connections and mobility between them (Landesregierung Nordrhein-Westfalen 1968, 48). Figure 3.1 shows how the region was conceived in terms of clearly distinguishable zones of work, home, transport, and leisure. The SVR's plan recommended the "unmixing" (*Entmischung*) and "densification" (*Verdichtung*) of residential and industrial areas and the use of green open space to "loosen" (SVR [1960/61] quoted in Benedict 2000, 88), differentiate, and bind these disparate zones and cities together. It proposed extending Schmidt's greenbelts with a planned "green open space system of seven contiguous greenbelts extending in a north-south direction" that would "take on a structuring function for the Ruhr," preventing sprawl, separating living and working activities (Pizonka 2010, 162), and "binding" and "rehabilitating" (*sanieren*) the "fragmented" (*zersplittert*) landscape (Benedict 2000, 88–89).

This principle of a functional division of urban space, refracted through local opportunities and concerns of the day, produced four new physical forms in the landscape. First, in the realm of transit, was the car. Hitler pushed mass motorization and highway construction in Germany in the 1930s, but it was not until the late 1950s that increasing personal wealth and the decreasing price of consumer goods made private automobile ownership possible for working-class families in Germany—something that had already occurred in the United States—and helped turn cars into symbols of freedom and independence. Throughout the Ruhr and West Germany, building the "automobile-oriented city" was a central goal of the postwar years (Landesregierung Nordrhein-Westfalen 1968; Reichow 1959). Second, in the world of work, was the university. Opening universities in the Ruhr was one response to the crisis of the industrial economy, and their appearance marked an economic shift toward light manufacturing and a service economy. The first opened in the Ruhr in 1965, and five more quickly followed. They were intended to combat the region's shrinking population and produce the first college-educated Ruhr generation, turning the children of miners and factory workers into an upwardly mobile "bourgeoisie through training" (Goch 2002a, 101, 104). Third, in the realm of domestic life, was the high-rise apartment block. Though the residential high-rise could sometimes be a symbol of "massification, isolation, and . . . decay," it also reflected new postwar domestic ideals (Herlyn 1970, 17) as domestic space became a place of retreat within the new functionally divided city. High-rise construction took place throughout the Federal Republic in the name of "modernization" or "rehabilitation." In the Ruhr, it was also part of an effort to address the "unsatisfactory housing structure" that was a result of "early, rapid, unplanned growth" by separating housing from industry (SVR 1966, 69).

Fourth and finally, planners found the public realm in green space. The SVR responded to the call for the public sector to "take on" the "leisure-time problem" by creating accessible facilities with the goal of "creating a citizenry" (Hezel 1974, 137) and by giving it a physical place, temporal location, and functional role in the oasis of the regional park. In the mid-1960s, "on the basis of scientific and practical advisory boards and research on leisure-time behavior," the SVR made "open space and leisure policy" (*Freiraum- und Freizeitpolitik*) its central focus (Grothe 2003, 39). With leisure time predicted to increase by one-fifth by 1980 (Landesregierung Nordrhein-Westfalen 1970, 107), it made the recognition of *Freizeit* "as an expansive sphere of society" one of its major long-term goals and promised to create, by 1975, "a single governing state structure for leisure and recreation" (Landesregierung Nordrhein-Westfalen

1970, 113). The SVR's "typology of leisure landscapes" in the Ruhr included a variety of indoor and outdoor leisure and recreation facilities, but the "heart of the concept was doubtless the *Revierpark* as *Freizeit* park" (Grothe 2003, 39, 40). As the region "broke up [its] large area into functional areas and structures" (Kastorff-Viehmann 1998, 131), the *Revierparks* were designed to be the physical space in which that leisure time could be productively spent.

The SVR understood the *Revierparks* as "completing" the existing network of green spaces to meet new needs and uses by being the first dedicated to leisure time recreation (Benedict 2000; SVR 1969, A24). In part, this new leisure infrastructure anticipated a transition to less physically demanding jobs—light manufacturing instead of coal and steel and an increasingly rationalized production process—that required not just passive rest but active recreation, sport, and play as *Erholung* (recovery) (Hezel 1974, 136; Lüdtke 1972, 39). Schmidt's greenbelts played a "structuring" role but were not for active use; patches of "'undisturbed' nature" on the region's outskirts provided destinations for weekend excursions but were "no longer sufficient to meet the new demands of relaxation and recreation" (Landesregierung Nordrhein-Westfalen 1970, 108), and the lawns of the existing city parks (*Stadtgärten*) were places for walking and taking in the flowers, not for laying on the lawn (Gaida and Grothe 1997, 28). The SVR also met its democratic mandate by locating the parks in the most densely populated areas of the region—on the borders between cities, accessible by public transit as well as private automobile (Landesregierung Nordrhein-Westfalen 1970)—to provide open spaces that working people could easily reach from their homes (Gaida and Grothe 1997, 29). *Revierparks* were, in a sense, the spatial analogue to citizen participation in the planning process. As participation was meant to secure a kind of procedural democracy, the parks—as the location of the nascent public realm—became settings in which to realize urban democratic ideals.

## REVIERPARKS AS THE SPACE-TIME OF DEMOCRACY

Having turned away from the ideals of public life and built form of the nineteenth-century cosmopolis, planners assigned the parks the traditional functions of a town square or marketplace: *Revierparks* were where democracy was to be enacted, in the context of new forms of anonymity and exposure to difference. For Ruhr planners, large parks in a functionally divided city represented Habermas's ideal in the landscape and, like Krupp's garden cities, were also understood to help create this imagined public sphere and its public. According to Lüdtke, the Ruhr's *Revierparks* furthered the goals of an "open and democratic" society by making the leisure sector an area of rich, diverse, and open exchange; by creating a context in which class differences would be "invisible"; by helping "values and goals that come out of leisure experience . . . become a meaningful source of dominant values and goals for the whole society"; and by facilitating the transformation of "emancipated consumer-citizens" into "emancipated political citizens" (Lüdtke 1972, 49–51).

The SVR made these intentions explicit in a 1977 brochure designed to introduce the new *Revierparks* to the German public upon their completion. Titled *Hier bin ich Mensch: Oasen einer Industrielandschaft* (Here I am human: Oases in an industrial landscape) (SVR 1977), the brochure served as an instruction manual for the parks' use by showing people how to behave in them and who to be in them. It also illustrates how planners intended the large, multiuse *Revierparks* to serve as a locus of democratic public life—to provide opportunities for "rational-critical debate" in a public setting (Habermas 1991a, 43) and to be spaces for learning how to participate in the new urban democratic society—and why and how green spaces were uniquely suited to that goal.

The brochure's title phrase introduces the parks as "oases" of leisure. "Hier bin ich Mensch" is a line from Goethe's *Faust* that is a popular and well-known saying in Germany. It comes from a passage called "Before the City Gate" that describes an Easter walk through the blossoming countryside, physically outside the city and temporally removed from the worries of work and daily life. The full sentence is often translated: "Here I am human, here I am allowed to be," and it is used to describe settings in which one feels comfortable, at ease, and fully oneself. In the context of the overall work, it is a moment when Faust ceases his striving and connects to simple earthly pleasures and a common humanity. In the context of a spatial division of functions, the value of *Revierparks* as oases of freedom from work and home was to create a place for Ruhr miners to be part of and to pass in the new middle-class society. Throughout the brochure, images of working men, retirees, and young couples lounging in grassy expanses emphasize the freedom of solitude in public space and describe how the parks offer privacy ("Mother can't see you, the children are far away"; they are an escape from the "madhouse" at home) as well as opportunities for public discourse. They allow working men to "explain to unknown mates how full of responsibility that shitty job is that you have to do," to "[look] for recognition, [hand] out recognition, [forget] all the tedium of the week, [forget] one's supervisor, who always wants . . . to bring [one] down a peg or two."

Far more than Metzendorf's plans, Osthaus's essays, or Krupp's deathbed wishes, in 1970 planners, policy makers, and public officials defined the Ruhr's present and aspirational future as a success in contrast to a negative or rejected past. This was a common narrative at the time. A book published by the city of Oberhausen in 1975, for instance, described two successive crises overcome. After the Second World War, "Oberhausen had been reduced to a heap of rubble by 161 bombings," but "ten years later all the furnaces were burning again, and 240,000 people had work, bread, and a home." "It happened for the second time in the 1960s and 1970s," the book continued, "when worldwide recessions caused structural changes." However, by 1975, "reconversion [was] in full swing," "new shores appear[ed] on the horizon," and the "industrial metropolis with its delicate one- or two-industry structure" was being prepared for a "more beautiful and secure future" (Buhrow and Holtappel 1975, 43).

The *Revierparks* brochure brings this hopeful, Habermasian future into focus by contrasting it to two things to be left behind: National Socialism and the industrial economy. National Socialism and its legacies are represented by the colonies and the industrial economy by the soot, dirt, and pollution to be replaced by green nature. The brochure repeatedly juxtaposes full-color, full-page spreads of the desirable democratic future against small black-and-white photographs of the industrial past. People sprawl on emerald lawns, play on new playgrounds, and hang out in diverse groups, while images most characteristic of the coal and steel economy—smokestacks against the sky, children playing in factory yards, and elderly people sitting on the steps of colony housing—occupy the edges of the frame. The "before" picture—of a provincial, polluted, industrial Ruhr—features colony life. The "after" picture—of the postwar democratic leisure economy—depicts the Ruhr as an enactment of Habermas's normative vision of the bourgeois public sphere: a public realm of rationally communicating, comfortably middle-class citizens.

As they did at Margarethenhöhe, planners had to envision the new urban public that was to be cultivated through green space, and once again that involved the transformation of working-class industrial workers into democratically oriented bourgeois citizens. Though the parks were part of the effort to attract new residents and new businesses to the Ruhr, as were the universities and the high-rise housing, they were also a social-managerial effort that was part of remaking the Ruhr's people to reflect new social, political, and economic ideals. But, in keeping with the '45ers' values of democratic management, and because planners were public servants rather than employers and therefore bound to the public interest, the only managerial style available to them was a cooperative one. The parks really were located to serve the Ruhr's working class (Benedict 2000, 98), and the brochure directly addresses those residents. In addition, the SVR gave the parks *volkstümliche*, or "folky," vernacular names: calling them *Revierparks* was a reference to the popular name for the Ruhr region (*Ruhrrevier*; Priebs 2019, 180–81), while the names of the individual parks referenced medieval villages now incorporated into the newer cities, old woodlands on which the parks were now located, and plants native to the region—all locally significant, though presumably only to longtime residents. While acknowledging this history, the brochure also introduces Ruhr residents to their future selves that are expected to emerge through these new experiences of work and leisure with a collage of smiling faces—a handsome young man, a pensive baby, dimpled boys, hardworking people—announcing:

> [The people of the Ruhr] no longer live in closed communities, in colonies: ghetto-like, the people on Rhine, Ruhr, Lippe, and Emscher. They spend their holidays in desirable scenic areas like other Germans. They are no longer the mining buddies [*Kumpels*] of the old days that others elsewhere—a little indulgently—laughed at. They have (perhaps) a somewhat more intimate relationship to work than their colleagues in other areas of Germany. But they know their own value, they have discovered themselves, they can no longer be so easily taken for a ride.... They know Bayern and the Adriatic; they feel as at home on the Italian Costa Watchamacallit as they would on the North or Baltic Sea [i.e., in Germany]. (SVR 1977)

In the context of adopting an urban-democratic model of a spatial division of functions and pluralistic leisure society, planners' rejection of the colonies was a way of discarding the insularity and homogeneity of community as a social form that had taken on new, exclusionary, and oppressive connotations in the shadow of National Socialism. In the passage quoted above, the past being left behind conjures the *Der Spiegel* article's image of noisy flocks of vacationing miners, reminding residents of other Germans' and the international community's perceptions of the Ruhr. Describing the colonies as "ghetto-like" evokes the racism and xenophobia of National Socialism that the Federal Republic was determined to overcome and associates that sentiment with the Ruhr's industrial workforce and company housing. Instead, a change is promised. Rather than harboring the false hope that "genuine democracy" could grow from "idealized closed neighborhoods," the *Revierparks* (along with highways, universities, high-rise housing, and new employment possibilities) fulfilled West German social scientists' recommendations of a "healthy kind of urbanity" where the "city air" of an open society "makes one free" and "people of all kinds of necessity come into contact with each other, communicate with each other, and accommodate each other" (Diefendorf 1999, 13–14). The *Revierparks* brochure promises that, as a result of such spaces and experiences, the Ruhr people will become worldly, well-traveled, and well-informed citizens, no longer naive and provincial and "indulgently laughed at" by outsiders.

Reflecting the political vision of the '45ers, the parks provided access to democracy and middle-class experiences through leisure time—a space that does not overcome class differences but that does create a context in which those differences can become "invisible"—rather than through upward social mobility. According to Lüdtke, parks did this by providing a space for social interactions not colored by work, by promoting leisure activities not based on consumption, and by offering a setting that minimized class location as a potential barrier to interaction. Lüdtke also believed in the broader benefit of these interactions, arguing: "[The] social integration that occurs in leisure time has consequences for the whole social structure in the sense that *the values and goals that come out of leisure experience become a meaningful source of dominant values and goals for the whole society*" (Lüdtke 1972, 49–50). Other planning and economic interventions in work, home, and transit might produce upward social mobility in material terms, but, because everyone participated in leisure time, it was argued, everyone could participate in democracy. And, to the extent that leisure was understood as the setting for democracy, it was through the parks that people were to become democratic citizens: worldly not provincial, savvy not superstitious, and discerning and self-possessed, part of or at least at home in comfortably middle-class West Germany.

The *Revierparks* brochure also provides a model of what these rational-democratic leisure interactions and Habermasian publics actually look like, bringing the future into focus by contrasting it to the past to be left behind. A typical spread shows interaction among strangers across gender, ethnicity, and class (fig. 3.2). On one page, women, photographed from above, sprawl on the ground with their children. An inset color photograph shows a card game in progress. A group of men play as women lean in and watch, their headscarves indicating that they are Turkish immigrants. An inset black-and-white photograph shows a nuclear family leaning out of a ground-level window of what appears to be colony housing. The text reads:

> Before, they leaned out the window, after work, on Saturday, on Sunday. Everyone knew everything about everybody: who with whom and when and where. They also knew where extra hands were needed, knew who was sick, who was pale with grief, knew where help was necessary, just stopped by.

FIGURE 3.2. Freedom as cultural pluralism. (From SVR 1977. Photograph by Manfred Ehrich, Essen, Germany.)

It's no longer so easy with neighborhoods, not so easily understandable [*überschaubar*] as in the times of closed society. It's perhaps more anonymous today, probably more intimate too. Better? Worse? Who can judge? Different in any case.

But you need "heat in the stable" if you haven't learned to be alone, you need others, you want to talk; without feedback, without resonance, you can't *be*. You seek contact, you want to communicate, without suspicion and shyness, just like that. And you understand the new ones, the other, the Turk, the Italian, who can't speak German so well yet.... Maybe you think that your grandfather.... [O]h, forget it—Come here, you! Do you play chess? Ball? My, you have such beautiful eyes. (SVR 1977)

This passage, and the depiction of Turkish and Italian immigrants, reflects the extent to which the regional parks were understood to be places for learning to participate in the new democratic society in concrete terms, rather than just representing an abstract commitment to diversity. The past to be escaped is the claustrophobia of *Gemeinschaft*—where "everyone knew everything about everybody"—represented by the colonies; the present that is "different" reflects the Ruhr's higher-than-national-average immigrant population, drawn by the region's constant need for industrial labor, its revived postwar production, and its low unemployment rate. Much as immigrant labor from Poland and the eastern provinces fueled nineteenth-century industrial growth, immigrants and refugees filled an industrial labor gap after World War II. Forced laborers made up roughly 40 percent of the Ruhr's World War II workforce (Kift 2013, 506), and, after they were freed, new miners had to be recruited—almost 400,000 between 1948 and 1953. In 1950, close to a quarter of the miners in the Ruhr had a refugee background (Kift 2005 cited in Kift 2013, 506; Kleinert 1988; Stahlberg 1957). Additionally, in the 1960s, available jobs drew about four million guest workers and their families to the Ruhr from Turkey, Italy, Greece, and Yugoslavia (von Einem 1982, 15). And so, in the 1970s, the region still had an immigrant population that was higher than the national average.

The women and immigrants in the scene described above anticipate the later critique that not all could participate equally in Habermas's brand of communicative rationality—but, for precisely the same reasons, the image reflects Habermasian ideals. Understood as zones for interaction among strangers across ethnicity and class, the parks are shown allowing women, children, and the Ruhr's ethnic German and immigrant working class to come together freely as strangers in public, without fear, and without discrimination on the basis of class, gender, or ethnicity.

## EXPERIENCING GREEN AS GOOD

Leisure time was one aspect of the parks as a democratic, universally accessible public sphere; the phenomenology of nature experience was another. The *Revierparks* brochure rapturously depicts pleasurable sensory experiences of nature—images of swans on water, descriptions of the feel of grass between the fingers, or the sound of leaves rustling—and these highlight another aspect of parks' special social role as democratic spaces in a functionally divided city. As it represented one thing to be left behind—National Socialism—through the colonies, it depicted another—the industrial economy—as an environmental bad in the form of soot and pollution. In a context in which there were bound to be winners and losers in structural change, *Revierparks*' pleasant experiences of nature were physical referents against which the decline of industry could be recast as environmental gains for all. In addition to assigning civic purposes to the "oasis" of leisure time, planners also expected the new green parks and clean air to allow residents to soak up structural change as a positive experience that had produced a cleaner landscape and an improved quality of life rather than unsettling changes and an uncertain future.

FIGURE 3.3. *Revierparks* as structural change. (From SVR 1977. Photograph by Manfred Ehrich, Essen, Germany.)

This helps explain why the first spread in the *Revierparks* brochure is not, as one might expect, a picture of completed parks in use but an image of structural change (fig. 3.3). The only photograph with no people, it shows a grassy landscape freshly planted with young trees staked in still-bare soil, sculpted with gentle hills, cut by a sidewalk. The sky is white; the field is empty. The image suggests a blank slate: reality is about to change. The small, inset black-and-white photograph reminds us what it is changing from: children are playing and eating popsicles between a row of prewar cottages on the right and hastily constructed postwar apartment blocks on the left. The accompanying text reads:

> Sometimes shocks are beneficial, inspire the imagination, make you think, give new impetus. Sometimes ideas change reality.
> The big shock—for the people in the Rhine and Ruhr, was the [hitherto unthinkable] crisis of the mining industry.... Uncertainty weighed on the mining towns, [and many] left, especially young citizens.
> Planners and politicians call the therapy *structural change*. Rather incomprehensible for those affected. Improve the environment, build five large parks along the Emscher River, the most disadvantaged zone in the region—this is the

Siedlungverband Ruhrkohlenbezirk's answer. The idea is called *Revierpark*. It changes reality. (SVR 1977)

In promising longtime and new residents alike inclusion in a new democratic society through leisure time, the authors of the brochure—like the planners themselves—walked a delicate line, anticipating the future while trying not to leave anyone behind. The *Revierparks*, like the bourgeois public sphere, offered a model of politics that bracketed work and home and, therefore, economic inequality. This was a dramatically new definition of politics in a region that had historically been defined by strikes, industrial labor, and class conflict—but also a practical one given the region's likely future. Deindustrialization did herald the "expansion of the social groups that are rather imprecisely labeled the middle classes" as the Ruhr's white-collar workforce grew (Goch 2002a, 95-96), and more people were living a materially middle-class lifestyle thanks to a rising baseline level of affluence and the decreasing price of consumer goods (Hobsbawm 1969 quoted in Jelin 1974, 13), but these were changes in overall workforce composition and did not represent the new employment of former coal miners and steelworkers. How not to alienate the Ruhr's longtime residents and how to find a place for them in the new economy were real and challenging questions.

As the passage quoted above reflects, the *Revierparks* were understood to play a role in this transformation by helping make structural change "comprehensible" and by providing phenomenal experiences of structural change as positive and as something that all could participate in. Deindustrialization presaged alienation and unemployment for the majority of the Ruhr's population—the working class, employed in coal and steel. Children might attend university, and newcomers might find jobs in new industries, but the "hitherto unthinkable" crisis of the mining industry also brought about a regional identity crisis and challenged residents' sense of self. Habermas's vision of a new middle class defined primarily in terms of lifestyle rather than income, work, or politics (Tenfelde 2000; see also Jelin 1974, 11-12; and Goch 2002a) suggested the possibility of a cultural experience of structural change—of becoming worldly citizens and increasing inclusivity through leisure time and anonymity. In short, the parks offered an environmental experience of structural change, of leaving behind dirty industry for the blue skies and green grass of a postindustrial economy. And the universal accessibility of these experiences contributed to the narrative of structural change as beneficial for everyone: the fact that everyone got an equal share of cleaner air was the primary way in which those who stood to benefit the least from structural change experienced those transformations as positive.

Experiences of structural change as environmental improvement were, objectively, real, beyond *Revierparks*, and not only in the 1960s. The SVR first voiced concerns about air pollution in the 1920s (Museum Folkwang 2010, 76–77) because, during the French occupation of the Ruhr, industrial production ceased and, "suddenly," "the air was of the same quality as in nonindustrial regions": "The leaves, which normally started to wither in early summer, stayed fresh and green until autumn. Potatoes grew bigger and fruit and vegetables, which normally were covered by a layer of dust, soot, and tar, remained clean and delicious. The harvest was approximately 50 percent higher. Even the tree rings grew wider than in the years before or after" (Bergerhoff 1928 quoted in Brüggemeier 1994, 41). The possibility of cleaner air was felt again in the 1940s, when plans to deindustrialize the Ruhr halted production after World War II (Brüggemeier 1994, 43). And, today, the story of washing on the line reported by *Der Spiegel*—that housewives could hang laundry out to dry only under favorable wind conditions—has become apocryphal. When asked about deindustrialization, Ruhr residents inevitably begin by noting that structural change has overall been positive because, while their mother's, grandmother's, or aunt's laundry turned gray from ash within minutes, today the air and the laundry are clean, plants and animals have returned, and one can enjoy being outside.

Greening efforts throughout the region helped citizens experience the postindustrial Ruhr as a clean, desirable place in the 1960s and 1970s. Ecologically oriented "hygienic" greening to help clear the air included restoring woodlands, planting trees and flowers, preserving and continuing to expand green corridors between cities and around the perimeter of the Ruhr (Benedict 2000, 87–89; Kastorff-Viehmann 1998, 131), and greening spoil heaps (*Halde*), the many mountains of extracted mining waste (Knabe, Mellinghoff, Meyer, and Schmidt-Lorenz 1968; SVR 1974). These activities were carried out in the name of public health and in response to both local environmental conditions and the concerns of the burgeoning environmental movement. Nationally, the planning profession was developing its green consciousness, in response to reports such as those in *Der Spiegel* on toxic environmental conditions and escalating concerns about air and water pollution, dying forests, scarce energy supplies, and nuclear power (Engelke 2011). Brandt's "blue skies" speech also amassed surprising support. The theme was taken up by the Christian Democratic Union after Brandt's election bid failed, and, locally, it helped rally residents to fight for legislation that gave citizens rights to demand that industry take steps to curb pollution ("Zu blauen Himmeln" 1961, 25).

A second set of green enhancements was cosmetic and directed toward outsiders: "landscape beautification" (*Landschaftliche Verschönerungen*) efforts focused on leisure and transportation amenities. These were a weapon in the "battle against the coal-pott* image" (Benedict 2000, 84, 103), geared toward improving the region's appeal as a commercial and residential destination for new people and businesses, hiding evidence of the struggling industrial economy, and facilitating a pleasant experience of the landscape on foot, by car, or by train. As the NRW put it in 1968: "The inner Ruhr area has more green space than is commonly assumed. As long as these areas are not used for agriculture or forestry or maintained as gardens, their recreational value is low. These often fallow and untended areas, on which old or unused equipment stands, greatly contribute to the negative impression that parts of the region evoke in strangers to the Ruhr. This impression is reinforced by the often untidy and monotone exterior of factories and residential and commercial buildings" (Landesregierung Nordrhein-Westfalen 1968, 67).

The SVR and the NRW began hiding "untidy and monotone" factories and commercial buildings with green in the 1950s by planting along the edges of railways, roadways, and parking lots. "Planting with the goal of screening" (*Abschirmung*) meant that factories and collieries that had once been celebrated as monuments to the Ruhr's economic power, technological sophistication, and political influence were now hidden; factories near railways, such as the Zeche Zollverein (which would return to visibility as a museum at the end of the century), were shrouded in "green curtains" to obscure them from the view of passing trains (Benedict 2000, 84). By the 1960s, bike paths and trails were also being greened as the SVR's 1966 development plan aimed to turn existing "recreational infrastructure" into new, enhanced, "leisure time opportunities" (Benedict 2000, 85; see also SVR 1966). Over ten years, the SVR installed five thousand benches throughout the region, in cities as well as in the towns in rural areas (Benedict 2000, 85). And, in 1970, the NRW added to its goals the "beautification of the urban landscape," describing activities including planting trees and flowers in city centers and greening coal heaps to make them more "eye-catching" (Landesregierung Nordrhein-Westfalen 1970, 111–12).

Planners and politicians assumed that *Revierparks*, green corridors, and new benches were amenities that the Ruhr's new postindustrial workforce

---

* A sometimes-affectionate nickname for the Ruhr, one foregrounding its industrial character.

would expect and require. At the Second German Leisure Congress in September 1972, the state secretary of the Ministry of the Interior announced that regional and local governments had had the "insight that the recreational value of a city and a region often becomes the decisive factor in selecting a residential or commercial location" (Grothe 2003, 39–40). The NRW's 1975 plan also states: "Leisure and cultural life are gaining in importance. . . . Whereas a few years ago employment and money-earning possibilities were the sole criteria for migration, today recreational value [*Freizeitwert*] is ever more at the forefront" (Landesregierung Nordrhein-Westfalen 1970, 107). City and regional marketing promoted this message—and these new experiences of the Ruhr—in the same decades. The Ruhr city of Dortmund, to which we will turn in part 3, began a marketing campaign promoting itself as "49% green" (Historischen Vereins für Dortmund 2012, 6) and organized city tours and urban hikes through the "blue cloud landscape" (Historischen Vereins für Dortmund 2012, 6; Benedict 2000, 104). All these activities emphasized amenities such as the clear water, healthy woodlands, and clean air in the Ruhr.

### GREENING'S DEMOCRATIC OPTIMISM

Marshall Berman begins his classic analysis of social and economic modernism, *All That Is Solid Melts into Air* (1983), with a close reading of *Faust* as a "tragedy" of modernization understood as both cultural and economic development. His is not an explicitly urban reading, though his examples of midcentury "Faustian development" are the large-scale construction projects characteristic of postwar planning. He notes that Faust experienced a revival in the 1960s and 1970s, first as a pastoral character in the context of 1960s visions of "stable abundance, leisure, and well-being," and then as a villain once that postwar promise was broken (Berman 1988, 40, 82). When the *Revierparks* brochure was published with its Faustian title in 1978, West Germany was struggling with just this conflict. The invocation of the pastoral Faust reflects the '45ers' cautious hope for the future—the Fordist promise of comfort and quality of life, renewed democracy and a public realm.

As West Germans rebuilt public life after World War II, *urbanity* was the word that scholars and planners used to describe this civic aspiration at the municipal and the regional scale—a version of it based on sharp distinctions between public and private and a bourgeois public characterized by their relative anonymity, rational discourse, and shared experiences of leisure. With that goal in mind, green space became a valuable planning tool. As part of this modernist planning paradigm, *Revierparks* represented a postwar democracy in which everyone was supposed to be able to participate, an urban future in which everyone was included. And green public spaces were not perceived to function, socially and politically, as nationalist or antiurban returns to nature, not *escapes* from society, but as postwar centers of modern democratic leisure that were a defined and distinct part of society. They were to serve as the locus of public life and, through the interactions that took place within them, shore up beliefs, values, and behaviors that would benefit the public sphere.

The *Revierparks* brochure's invocation of the pastoral Faust is also evidence of why nature's status as a universal public good might create an affinity between green space and urban democracy. For Faust, the utility of a space perceived to be free is as a moment of freedom and relaxation, in which he got to enjoy being human. For park planners, the same qualities gave nature political and social functions as a place from which all would benefit and in which strangers would come together and enjoy a shared experience of common humanity. What better setting than that could there be for deliberative democracy? Nature's ideology of universality made it a potent vehicle for this message. Ruhr planners met their public mandate by using nature to carve out a new urban public realm through parks that helped define a new middle class in terms of leisure rather than work and consumption in terms of light, air, and green space rather than purchasing power. This made it possible for the public sector to narrate a story of structural change as good for everyone and sidestepped discussions about the new inequalities deindustrialization was likely to produce.

Through the *Revierparks* we can also begin to establish continuities in greening's logics across periods as the dynamics introduced in chapter 2 reappear here in the context of a very different political economy and power relations. Regarding greening's compatibility with modern liberalism, its utility as a social-managerial technology continued and became even more important in the postwar era. Habermas begins *The Structural Transformation of the Public Sphere* by arguing that, since the creation of the liberal public sphere, publics had diversified, raising the question his book tried to answer—how to create a public sphere for rational debate among diverse publics. For Ruhr planners, green urban public space turned out to be uniquely suitable for this purpose because the parks, and leisure time, appeared to offer universal accessibility indifferent to class, ethnicity, or employment status. Paternalism also reappears—in the form of good intentions combined with social-managerial goals—even in a context in which greening protagonists were public servants rather than employers with direct control over the users of green spaces.

Ruhr planners understood this technocratic approach to democracy through leisure time inclusion as pragmatic rather than naive or utopian. The

'45ers were the "skeptical generation" (Schelsky 1963), well aware of the potential tragedies of development. Edgar Salin expressed doubt about attempts to revive urbanity in the context of capitalism and mass culture and was critical of policy makers' failure, when establishing the new Federal Republic, to fundamentally rethink nineteenth-century concepts—such as that of property—that would have "emancipated" people from feudalism and made a true "society of urban citizens" (Salin 1970, 877; see also Salin 1960, 31). The NRW's planning documents acknowledged: "The responsibilities of a government do not allow it, in its programs, to paint a utopian *Fata Morgana* [i.e., a mirage] of wishes on the horizon of development. . . . We will not realize utopia in the 1970s" (Landesregierung Nordrhein-Westfalen 1970, n.p. [foreword]). Yet the *Revierparks* brochure is cautiously optimistic about the parks' ability to affect change. On the back, an image of children playing high on a rope climbing gym against a smoky, ashen sky features a quote from Manès Sperber, a French-Austrian novelist and psychologist whose work was well-known in West Germany: "I am an opponent of utopias that promise what they could never fulfill. Yet I stay in the vicinity of utopians because their enthusiasm sometimes helps gradually to effect those changes." This reflects planners' and scholars' attitudes toward the parks: as a pragmatic means of establishing a new model of urbanity and democracy, a foundation for rebuilding national community on "social-psychological rather than national-cultural criteria" (Koshar 1998, 277).

Beyond the brochure, sociologists of the postwar era also described public space in these terms. In her defense of the high-rise apartment block—perhaps the most controversial part of modernist planning's functional division of spaces—the sociologist Ulfert Herlyn argued that it is only when viewed in isolation that the high-rise looks forbidding. Viewed according to the "principle of functional division," it is not an alienating symbol of "massification" but a comforting and necessary space designated for privacy and retreat. It is even "emancipatory." Herlyn argued that shorter workdays and a well-defined domestic realm would help "instrumentalize" work—constraining the space and time for earning money—and "liberate" public spaces and leisure time for play and civic life (Herlyn 1970, 29). Bahrdt argued that, as part of a specifically *urban* life, the home as private retreat preserved the "tense interplay" between public and private "in such a way as to shape a public and a private sphere that stand in relation to each other in a permanent tension and complementary relationship, and through which give urban life a special dynamic, generosity, and freedom—in short, the urbanity—never seen in a closed system," while allowing the "extremely sensitive and intimate areas of [private] life" to be better protected and to flourish (Bahrdt 1952, 1474, 1477).

Of course, not all agreed that the version of democratic urban public life achieved in these spaces was desirable. As critics of Habermas would later argue, urban public spaces were no more universally accessible than was the ideal of deliberative democracy itself. And it was exactly these modernist planning principles that would eventually be blamed for the decline of urban life in Germany and the United States—described retrospectively as the destruction of the urban public realm (Lofland 1989, 22) rather than the creation of a new one. But, at the time, planners understood themselves to be creating a new, universally accessible urban public life through a functionally divided landscape and large-scale, anonymous public spaces. As part of this vision, *Revierparks* represented philosophical and political ideals of diversity and anonymity, a pluralistic public sphere and the democratic common good.

CHAPTER 4

# Proletarian Counterpublics: Reimagining the Colonies

In 1972, one of the remaining steel companies in the Ruhr proposed what should have been a noncontentious act: demolishing the region's first and oldest workers' colony, Eisenheim, built in 1844 and located in Oberhausen, a town northwest of Essen on the western edge of the Ruhr. The colony was in a state of physical disrepair, and this model of housing had fallen out of favor. And so the mining company that owned it (August Thyssen Hütte, or ATH) proposed a plan well in keeping with the modernist planning and democratic political ideals that crystallized in the *Revierparks*. It would "rehabilitate" (*sanieren*) or "modernize" (*modernisieren*) Eisenheim by replacing its 39 two-story houses—containing 139 residential units and housing just under 500 residents—with 400 high-rise apartment units (Boström and Günter 1976, 43; Günter and Weisser 1975, 92, 93). But ATH was surprised to find itself facing strong and organized resistance in the form of a "workers' initiative" (*Arbeiterinitiative*) that organized to fight the demolition of the settlement—and won. Following the movement's success, about fifty more *Arbeiterinitiativen* organized to save other threatened colonies in the Ruhr, using Eisenheim's tactics (Faecke, Stefaniak, and Haag 1977, 79; Mesecke 2010).*

---

* Estimates of the number of workers' colonies remaining in the Ruhr in the 1960s vary widely. Even when colonies are defined precisely—as "relatively closed residential housing units with an essentially village-like character, built by mine owners since the middle of the nineteenth century for their own workers"—counts range from 700 to 3,000 settlements, containing from 350,000 to over 1 million residents, though the median is around 2,000 (Nelles and Oppermann 1979, 49).

Much of the Eisenheim movement's visibility and success was due to the efforts of Roland Günter, the son of a factory worker and a professor in nearby Bielefeld who still lives in Eisenheim and maintains a library and archive of the movement on-site today. When Günter and a colleague brought a group of students to study the colony in the early 1970s, they "discovered" an "intact social fabric" (*Sozialgefüge*) (Projektgruppe Eisenheim 1973, 59) that they wished to recognize and preserve. Over several years, Günter, his wife Janne, and his students worked alongside Eisenheim's residents to save the colony through media campaigns, local organizing, and political advocacy (Nelles and Oppermann 1979, 57), in the process producing two books and a variety of public relations materials that this chapter analyzes. In direct contrast to the democratic ideals of bourgeois rationality embodied by postwar high-rise housing and the *Revierparks*, and drawing on the work of two of Habermas's students, Oskar Negt and Alexander Kluge, the movement's language and materials celebrated the colonies as fostering a "half-public," "proletarian," or—as we might say today—"counterpublic" sphere (Krause 2006; Warner 2002) of situated, classed, historical individuals. And it was nature—Eisenheim's flower gardens, kitchen gardens, chickens, pigeons, and shared green spaces—that were upheld as symbols of these values, and that the movement argued facilitated forms of familiarity, mutuality, and solidarity that would be lost if the colonies were destroyed.

As part of a swell of New Left social movements in the 1970s, the Eisenheim movement offered a second green answer to the question of how best to build democracy and urban public life in the postwar world. Disillusionment with modernist planning and the limits of legislated democracy had grown in the Ruhr—as it had across the West—in the 1960s, and the movement proposed a very different model of urban publics and democratic politics. Rather than a bourgeois public sphere founded on sharp distinctions between public and private that made class differences invisible, it advocated for a proletarian counterpublic sphere that made personal difference, history, and experience the foundation for public politics.

Paired with the *Revierparks*, the Eisenheim movement helps establish that urbanized nature is a social imaginary of form and not content, a way of expressing normative visions rather than any particular one. It does so in part simply by coexisting with the *Revierparks*, by showing that multiple social visions can be communicated through green space in the same place and time. But, because it is a Left vision, it also demonstrates that social visions with very different relationships to power can be expressed through nature in the same period. As in the *Revierparks*, nature was the medium for realizing this alternative urban vision, but, unlike the *Revierparks* planners, the movement understood the benefits of green space to be in alignment with more radically transformative political goals. The comparison of the two projects also highlights similarities in greening's dynamics in spite of their political differences, further substantiating the argument that greening projects play out with similar logics across changing time periods and very different relationships to power.

By documenting a case in which urbanized nature is used for the reinterpretation of existing natures rather than the design of new ones, this chapter also elaborates, empirically, a moment of cultural creativity in its material and historical context. Previous chapters have traced the emergence of new urban ideals through entirely new green constructions. Here, the Eisenheim movement provides an opportunity to explore *reimagining*—how a repressive/oppressive space came to be seen as a liberatory/progressive one in the context of a new political economy and why, how, and under what conditions the imaginative content of old forms can change. The Eisenheim movement, which based its arguments for preservation on the forms of sociality and political consciousness produced by the colonies' unique green spaces, understood themselves to have discovered a proletarian public sphere in an environment that had existed for the past hundred years and reenvisioned it as an emancipatory space. It was able to do so in spite of the colonies' associations with political conservatism, both in their nineteenth-century incarnation as company housing and in the context of negative postwar associations. Beyond the existence of urbanized nature itself, the factors that converged to make this new narrative possible included international social movements, national political trends, and local socioeconomic transformations.

## REIMAGINING NATURE: NEW ASSOCIATIONS WITH OLD GREEN FORMS

In 1973, a year into the campaign to save Eisenheim, the liberal national newspaper *Die Zeit* ran its weekly magazine with the cover story "Heimat oder Hochhaus?" (Homeland or high-rise?), which chronicled the debate surrounding the demolition. The article begins by describing the colonies in familiar terms and in keeping with dominant planning ideals, quoting the property owner and a regional government representative saying that colonies like Eisenheim are "simply proletarian . . . old junk" that should be "destroyed." But it is sympathetic to the preservation movement and goes on to quote residents flatly

110    *Contested Social Ideals*

FIGURE 4.1. *Die Zeit* depicts workers' culture positively and the "goods" of Eisenheim through nature. (From Schille and Rautert 1973. Photograph © Timm Rautert.)

rejecting the trade of "their familiar homes, their neighbors, their garden, and their animals" for "monotonous," "prison"-like new construction (Schille and Rautert 1973, 2).

In an almost direct inversion of the *Revierparks* brochure, which juxtaposes color photographs of *Revierparks* against the black-and-white past of the colonies, *Die Zeit*'s article also contains several large photographs showing the benefits of colony life—green space, gardens, and animals—against small, inset photographs depicting "barren," sterile high-rise living (fig. 4.1). "At third glance," in *Die Zeit*'s estimation (though at first glance in the eyes of the movement), rather than a "hole unworthy of human beings," Eisenheim began to appear as an "idyll" and "a green island in the black coal-pott." The article evaluates the colonies positively in comparison to high-rise living in large part by virtue of the access to nature they provide, explaining:

> In Eisenheim every family has a garden; they can raise chickens or care for pigeons. The pike man Kaspar Hasche even feeds five piglets that he slaughters on Christmas and Easter. And the kids can play where they want and make noise without a doorman chasing them. There are only a few fences, hardly any signs that read "Playing forbidden," and none that read "Do not step on the grass." The houses border a big garden area on which trees stand, flowers bloom, potatoes, kohlrabi, onions grow. When the men get off their shift, when they come up from the tunnels lying eight hundred meters underground or

leave the glowing furnaces, they go straight across the courtyard to their gardens. In Eisenheim there is always someone in the garden or in the courtyard or on the street. (Schille and Rautert 1973, 2, 6, 4)

As well as being evidence that this positive reinterpretation of Eisenheim transcended the movement and reached broader media, the inversion is remarkable for two additional reasons. First, rather than being a production of *new* natures, as we have seen so far, it reflects a reinterpretation and recasting of existing ones. Second, it reimagined those existing natures in a very different relationship to power. The Ruhr's company housing was an unlikely source of emancipatory consciousness, either as first constructed, as a tool of pacification and control, or as celebrated by National Socialists for its nativist, preindustrial feel (e.g., Heinrichsbauer 1936). For a century, the colonies had been symbols of working-class life, spaces understood to be synonymous with (at best) a kind of oppressive, feudal, paternalism and (at worst) insular and xenophobic National Socialism. In putting the colonies at the center of a radically different vision of democracy, the Eisenheim movement reinvented them in the service of Left politics.

How did this happen? How was the movement to save Eisenheim able to turn the colonies' pigs and potatoes from embarrassments or symbols of oppression into desirable spaces that could rally residents and make allies of mainstream media? The colonies were physically damaged in both world wars and devalued by postwar urban renewal and the ascendant leisure lifestyle that located the good life in the high-rise, car, university, and *Revierpark*. Recovering them as a favorable counterpoint to the high-rise and, even further, potentially seeing them as emancipatory spaces required an impressive rereading against this history.

Urbanized nature helps explain this transformation in two ways. It is, first, a way of explaining a common denominator: why both the Ruhr's urban planners and its New Left activists might have selected green space as a way to improve the region in the first place. In part 1, I argued that the classic understanding of greening as a reactive mode is incomplete by showing its aspirational use by Krupp, Osthaus, and Schmidt in their first efforts to make the Ruhr into a bourgeois city in the absence of the problems provoking greening elsewhere. Here, in part 2, this aspirational orientation reappears; both *Revierparks* planners and the Eisenheim movement were offering visions of alternative and understood to be normatively better social arrangements through nature. But, more generally, it helps explain how social imaginaries turn specific activities into recurrent "mode[s] of socio-historical doing" (Castoriadis

1997, 3)—or, in other words, how urbanized nature made greening a common way of representing social aspirations, a general, portable form of moral expression.

Second, the concept offers a response to the other possible understanding implicit in commonsense explanations of greening: that greening is an *ideological* reaction to the problems of the city, a distorted reflection of the social. Castoriadis's account of the social imaginary contests the idea of cultural and social practice as ideology by emphasizing the creativity of social practices— acts with the potential not just to reflect the world as it exists but to gesture at something new. Frustrated with deterministic variants of Marxism that he found inadequate for understanding social life and especially social change, Castoriadis echoed sentiments shared by cultural Marxists such as Lefebvre, Raymond Williams, and others who saw culture and everyday life as more than a "reflection" of material conditions. With the concept of the social imaginary, he was trying to explain the source of radical social imagination, mobilized deliberately to "create" a world (Castoriadis 1997, 3; see also Elliot 2004, 2012). He described the "self-creation of society" as a fundamentally creative, collective act, as "poiesis," as *making*, as "thoughtful doing" (Castoriadis 1997, 3, 5, 4), in order to emphasize not just how the social imaginary is productive but how it is capable of producing actions and institutions that are not predictable and sometimes fundamentally new.* But Castoriadis provided little information about what creative action actually looks like, in historical rather than theoretical terms. How is it that people actually engage in behavior that is not structurally determined or come up with new ideas within a given set of external, internal, and historical constraints (Gaonkar 2002, 7)? Under what conditions is this possible? What does it actually look like?

As a case in which the emphasis on the social and political power of urban green space coexisted with a Left structural analysis of the social and political problems of the present, the Eisenheim movement offers an opportunity to answer these questions. Even though the Ruhr's *Revierparks* and its garden cities were both aspirational projects and, thus, creative in that sense, both efforts to make the Ruhr into a bourgeois city were also system affirmative. But the Eisenheim movement chose a medium commonly understood as reformist and

---

* Social theorists have since extended this idea of imaginative work to a variety of social spheres, describing the "cultural creativity" that is involved in the construction of nations, states, cities, corporations, and other social institutions (Calhoun 2002) and part of the imaginative, agentic, "projective element" of response to problems posed by historical situation (Emirbayer and Mische 1998).

ideological and a place synonymous with exploitative social relations, through which to offer a radical, emancipatory vision through green space.

Within the specific historical context of the Ruhr, this chapter outlines the material and cultural factors that, at this moment of urban crisis, made the colonies available to be reimagined as nature and contributed to their being reimagined in these ways. As part 1 demonstrated, the capacity to imagine and act through nature at all is linked, spatially and temporally, to the modern urban era. So the first condition of possibility was urbanized nature's imaginative availability, which in the Ruhr can be dated to about 1910. The second was another "unsettled" time, a moment of urban restructuring when the Ruhr's societal norms and expectations and its political economy were changing, when international structural conditions, national cultural and political conventions, and local historical contingencies all influenced what could be "discovered" and revalued as "nature" and how those creative visions were articulated. Deindustrialization and postwar political concerns set the stage for the Eisenheim movement by producing locally influential international social and political movements, a new celebration of working-class culture, and the bankruptcy and consolidation of the industrial companies, the disappearance of which left company housing open to new forms of tenancy and social life. These were the factors that made it possible to reimagine once-oppressive company housing as a haven for a proletarian politics.

## BOURGEOIS DEMOCRACY'S DISCONTENTS

The global historical circumstances that provoked the reenvisioning of the colonies were a growing dissatisfaction with the bourgeois postwar political and consumer culture represented by Habermas and the *Revierparks*, and 1960s New Left social movements' disillusionment with postwar promises and very different vision of democratic politics. The leaders of the Eisenheim movement—members of the '68ers, a New Left generation composed of scholarly researchers, social movements, university students, and urban professionals that the '45 generation had helped birth, intellectually as well as biologically—had far less patience for the '45ers' pragmatic approach to democracy and inequality and the promise of social and political inclusion through leisure time. The Eisenheim movement's critique of the bourgeois model of democratic politics targeted the high-rise apartment block, calling it out as an empty signifier of social mobility for tenuously employed blue-collar workers: "Since this climb is not even really possible for most workers (in addition to which, the emancipation of the workers must take place by other means), the bourgeois high-rise apartment remains only an empty prestige signal—comparable to the chrome-plated car" (Projektgruppe Eisenheim 1973, 139).

In this sense, the Eisenheim movement should be understood as part of a wave of protest-based social and political movements sweeping Western Europe and the world in the late 1960s, swelling in the disillusionment of postwar promises, and offering a second answer to the question of rebuilding democratic public life in West Germany. In the United States and Western Europe, such social movements were associated with hippies, sex, drugs, and rock and roll—antiestablishment, nonviolent, anticonsumerist resistance to middle-class culture. In the Ruhr, and surrounding Eisenheim's demolition, the movement focused its critique on the vision of urban democratic society that we saw created in the last chapter. Dissatisfied with reformist democratic politics, and disappointed with the modernist planning paradigm, it railed against the functionally divided city and bourgeois model of democratic politics, offering an alternative to the *Revierparks*/high-rise model of urbanism and its vision of political culture.

In West Germany, the '68ers criticized Habermas—and initial postwar democracy more generally—for being too conservative and for producing "dutiful democrats" rather than active citizens (Roseman 1992 quoted in Prowe 2001, 451). They were also concerned that loyalty to democracy in principle was "leading to antidemocratic behavior" in practice (Thomas 2003, 3), such as the policing of student protest, tight conservative control of newspapers and other media, and legislation that allowed the government to suspend civil liberties in states of emergency (Flagge 1999, 860; Thomas 2003, 3). People were also frustrated with the growing conservatism of the Social Democratic Party (SPD), the traditional workers' party, as it rose in national power. Dissatisfaction swelled into a second movement of social and cultural democratization in West Germany. In the political space opened up by the efforts of the '45ers, grassroots, oppositional, movement-based politics grew (Siegfried 2005, 727; Von Hodenberg 2006). Along with "peace, ecology, Third World solidarity, squatters', women's, gay, and student movements (including the radical Communist and terrorist groups that sprang from the students movement)" (Koopmans 1995), citizens' movements offered radical new forms of "democracy from below" "aimed precisely against the hollowness of the new culture of public civility the 1945ers were so assiduously trying to produce" (Eley 2002, 228).

The '68ers' strong local presence was a direct result of deindustrialization. The flurry of university construction had changed the working-class Ruhr's socioeconomic composition by drawing faculty to the region and producing its first college-educated generation. This new "bourgeoisie through education" staffed and studied at the new colleges and universities and were active members of the '68 movements. In the Ruhr, they organized "citizens' initiatives" around quality-of-life issues stemming from urban renewal and economic development such as pollution, construction, noise, traffic, and energy issues as well as housing, tenants' rights, schools, and social services (Markham 2005, 670), and they also initiated "go-ins" (a tactic borrowed from American sit-ins), mass protest meetings, and takeovers of academic buildings (Thomas 2003, 131–32). In 1968 Essen was the home of the Internationale Essener Songtage, a five-day musical festival and "psychedelic happening" linking protest and popular culture that drew over forty thousand people (Brown 2013, 155).

Meanwhile, critiques of urban modernism were coalescing in America. From the failure of St. Louis's celebrated Pruitt-Igoe complex to New York City's devastated South Bronx, the promises of high-rise living and functionalist planning were falling short as people watched the social and physical decay that followed slum clearance, new highways, and large-scale construction. The architectural critic Charles Jencks famously declared the death of modern architecture "in St. Louis, Missouri on July 15, 1972, at 3:32pm," with the demolition of Pruitt-Igoe (Jencks 1977, 9). American sociologists began to critique the myth of "suburbia as a way of life" (Berger 1961, 38) and the desires of the new middle class: its consumerism, its social isolation (Friedan 1963), and the conformity and banality of its mass culture (Whyte 1956). In Germany, people were learning that they did not enjoy inhabiting dense, car-oriented cities (Engelke 2011, 39; Juckel 1992) or the air and groundwater pollution caused by rapid development (Brüggemeier and Rommelspacher 1992). In the Ruhr, where the speed of growth exceeded even that of the industrial era (Brüggemeier and Rommelspacher 1992), the real "social costs and ecological consequences" of highway expansion, suburbanization, and urban development were becoming apparent (Goch 2002b, 239). Sociologists and social psychologists also painted a dark picture of the effects of postwar planning and consumerism on individuals. They described "green widows"—wives isolated in high-rise towers all day while their striving husbands were at work (Dorsch 1974)—and the anonymity and loneliness of high-rise living (Flagge 1999, 861). Increasingly, concerned citizens returned to sociologists' warnings, such as Mitscherlich's 1965 *The Inhospitality of Our Cities*—the subtitle of which is *Instigation to Restlessness (Anstiftung zum Unfrieden)*—and used their critiques of prosperity, consumption, and freedom to "[oppose] urban renewal plans, construction, and road-building projects" (Roth 1991, 77).

The '68ers also brought a new interest in workers' culture to intellectual and artistic movements. The postwar Ruhr still had no distinct high "cultural identity" (Kift 2013), but rather a working-class reputation seen as antithetical to high culture in the bourgeois sense. As students and intellectuals became disillusioned with the '45ers' bourgeois vision of democracy and politics, recognition of the cultural value of industrial life and working-class consciousness grew. Art, theater, and literature helped "[develop] a new working class culture which did not have 'the movement' but the mining industry as its reference point" (Kift 2013, 503). The director of the mining museum in Bochum (one of the Ruhr's industrial towns), for example, wished to foster an "artistic engagement with mining" in the "still young and relatively uncultured coal mining area of the Ruhr" by having miners "paint their world" (Tenfelde and Urban 2010, 1003, 1025). In the 1950s and 1960s, sympathetic depictions of miners in pop culture also increased, such as the "pragmatic and unpretentious" fictitious miner *Kumpel* ("Buddy") Anton, who had a popular column about daily life and world affairs in the *Westdeutsche Allgemeine Zeitung*, a regional newspaper published in Essen (Tenfelde and Urban 2010, 1001). This paralleled growing scholarly interest in social history from below and the history of everyday life popularized by scholars such as E. P. Thompson ([1963] 2016), Raymond Williams (1958), Henri Lefebvre ([1947] 1991, [1961] 2002, [1981] 2005), and the Birmingham school. Among German scholars, the Ruhr became, by virtue of its workers' culture, a favorite site for this kind of research in the 1970s (Brüggemeier 1983; Lüdtke 1995; Niethammer 1983; Ritter 1978).

All these movements were part of a post-postwar New Left that was exchanging bourgeois for self-consciously proletarian ideals: from legislated democracy to grassroots politics, from high culture to working-class culture, and from modernist planning with its notions of privacy and anonymity to more solidary forms of community and neighborhood. They also identified a different locus of transformative social change, returning to the proletariat as the revolutionary subject and everyday life as the setting in which to create the conditions for radical politics.

This was the political context for the Eisenheim movement, which framed its critique of the high-rise/*Revierparks* model of urbanism and democracy in these terms. The stated objective of a movement-produced book titled *Rettet Eisenheim* (Save Eisenheim) was to "show the tight connection between the economic and ideological motives of the *Sanierung* [modernization]," and the book critiqued both the class relations and the ideology of individualism underpinning the proposed demolition. It also countered arguments based on material necessity, arguing that there was no actual need for new housing,* but that industrialists had, instead, declared the old housing obsolete to support new construction that would increase rental income. The movement also critiqued liberal and mainstream social scientists and planners:

> Hans-Paul Bahrdt . . . and others call for residential developments within which social ties are exclusively freely chosen and can be broken off at any time. . . . This reflects the extreme subjectivity of the individual, one who believes himself to be autonomous and fears that through social ties he will only be disturbed. Bahrdt—although he is against the single-family house—is the refined ideologist of the bourgeois "my house is my castle" (a phrase that could come only from the social isolation of the middle class in England). . . . As long as Bahrdt's claims remain limited to the bourgeoisie, they are understandable: it [the middle class] wears itself down by marital competition so that it makes sense that it wants to maintain the illusion, at least in its free time, that it is not tyrannized by competition. The bourgeois ideology that the apartment should serve the complete retreat of the self is an expression of bourgeois fears, which are heavily promoted through competition and the destruction of the protective social group in favor of "lonely" individuals. (Projektgruppe Eisenheim 1973, 3, 138)

### THE PROLETARIAN PUBLIC SPHERE

As Habermas inspired the construction of *Revierparks* as a bourgeois public sphere, the Eisenheim movement's alternative political vision for the colonies found a social-theoretical reference point in the work of two of Habermas's students, Oskar Negt and Alexander Kluge. Their *Public Sphere and Experience (Öffentlichkeit und Erfahrung)* was first published in German in 1972, the same year the movement to save Eisenheim began, and following publications by Negt on workers' consciousness and culture in the late 1960s. Though never as well-known as Habermas's text, particularly among American social scientists, *Public Sphere and Experience* was one of a number of sympathetic critiques of Habermas's bourgeois public sphere, offering an alternative

---

* There had been a huge amount of construction in the postwar, precoal crisis years. Between 1951 and 1959, ninety square kilometers of mostly agricultural land had been newly developed for industry, transit, and housing (Brüggemeier and Rommelspacher 1992, 87).

"proletarian" concept of the public sphere that extended and complicated Habermas's framework.*

Like Habermas, Negt and Kluge offer both a clear articulation of the normative ideal of urban publics and publicness that was guiding the Eisenheim movement and a theory of politics and social change that grew out of the West German experience in the same era. Like Castoriadis, they were interested in the genesis of potentially world-changing forms of consciousness and politics and in the radical potential of particular media to be put to use for emancipatory ends. Their focus was on the proliferating mass media, not nature, but their account of it makes a similar kind of argument: they believed that, in spite of its familiar stultifying and depoliticizing effects, it should also be possible for people to use the mass media to express emancipatory and counterhegemonic visions—in other words, as a *medium*, a flexible form rather than a specific (system-affirmative) content. They had their own language to describe this creative capacity of human beings; rather than social imaginaries, they speak of *Phantasie*. This word is usually translated as *imagination*, though sometimes, especially in Negt and Kluge's texts, it becomes the English cognate *fantasy*. Like Castoriadis, they take fantasy not as a utopian "expression" of alienation but as a mode of exploring alternative worlds and ways of being, an "unconscious practical critique of alienation" and, therefore, one of the "raw materials of class consciousness" (Negt and Kluge [1972b] 1993, 33, 35).

*Public Sphere and Experience* reflected the political frustrations of the '68 movement in an academic register. Just as demonstrations and direct action "exploded the terms of established political discourse," "theoretically speaking . . . [they] expose[d] some crucial difficulties at the center of Habermas's concept of the public sphere" (Eley 2002, 225). As social movements criticized bourgeois forms of democratic politics, Negt and Kluge criticized Habermas's bourgeois public sphere for its exclusion of proletarian publics, inadequate grasp of everyday life, exaggeration of emancipatory potential, and lack of attention to culture and identity (Calhoun 1992, 5, 34). Centrally, they exposed difficulties regarding the separation of private life and personal experience from public politics. They argued that, while the bourgeois public sphere "allegedly represents the totality of society," it excludes many people and experiences. By its definition, "federal elections, Olympic ceremonies, the actions

---

* Though primarily focused on the working class, Negt and Kluge also helped pave the way for recognizing more heterogeneous public spheres and communicative cultures in public space and public politics, anticipating feminist and race-based critiques (e.g., Fraser 1990; and Young 1990).

of a unit of sharpshooters, a theatre premiere—all count as public events." However: "Other events of overwhelming public significance, such as child-rearing, factory work, and watching television within one's own four walls, are considered private. The real social experiences of human beings, produced in everyday life and work, cut across such divisions" (Negt and Kluge [1972b] 1993, 63, xliii). These, the first sentences of the book's introduction, show how Negt and Kluge understood the concept of the bourgeois public sphere to exclude the central life experiences of the proletariat from its definition of politics. The types of experiences it deemed private—child rearing, factory work, watching television—were exactly the experiences that were drawn into domestic invisibility by functionalist planning and exactly those most significant to and characteristic of the Ruhr's working class. Their alternative concept of a proletarian public sphere was motivated by a vision of political participation grounded in concrete, historical individuals. Making *experience* a central analytic category, as the title of the book suggests, signaled their project to make daily life the basis for political participation.

Though direct references to Negt and Kluge do not appear in the movement's first published materials, which were written before *Public Sphere and Experience* came out, they do describe a "half-public sphere" (*Halböffentlichkeit*) and a "workers' public sphere" (*Arbeiteröffentlichkeit*) (Boström and Günter 1976, 67, 8). And Janne Günter's (1980) *Leben in Eisenheim* (Life in Eisenheim), written after *Public Sphere and Experience*, begins with the following epigraph: "The proletarian public sphere, emerging through the use of its organizational structures (forms), not only binds together proletarian interests and experiences, but also focuses on a specific stage of the proletarian public, which also formally differs from the bourgeois public—worker's club, workers' settlement, union" (Negt and Kluge 1972a quoted in Günter 1980, 13). Each of these versions of the public sphere is a counterpoint to Habermas's: "half-public" rather than a stark division between public and private space; oriented toward workers and, thus, class positions as lived and experienced rather than abstract unifying discourse; and proletarian rather than bourgeois. Negt and Kluge further bring the alternative model of democratic life motivating the Eisenheim movement into focus through the new ideal forms of politics, subjects, and urban space they describe.

First, they outlined an alternative democratic political ideal based on a principle of "mixture" (*Mischung*) rather than "division" (*Trennung*) (Jameson 1988, 160). Habermas's bourgeois public sphere divided public space from private, public personas from individuals' particular "life context," and individual interests from ostensibly universal ones. This division was reflected in the stylized forms of speech and communication that Habermas prized and, concretely and historically, in the spatial division of functions that produced spaces like *Revierparks*. Instead, Negt and Kluge argued for inhabited spaces and intact historical experience—characteristic of multigenerational, multiuse spaces and communities just like Eisenheim—that would draw on a "solid basis of real mass experience" to give "weight" to a proletarian public sphere whose interests were reflexively classed, historicized, and located (Negt and Kluge [1972b] 1993, xlvi, 80).

They also provided an alternative ideal of a political subject as a blue-collar producer rather than a bourgeois consumer. Negt and Kluge and the Eisenheim movement both worried about the depoliticization of the working class. Movement materials noted that "a class with a disturbed historical consciousness is obstructed in its political action" (Projektgruppe Eisenheim 1973, 3) and saw the production of individualized, consumption-oriented subjects as the way to such forms of pacification and social control. *Experience* offered a form of resistance and reintegration for fragmented, alienated subjects, a way to begin "the working through of the suppressed experience of the entire labor movement that has been mutilated by the bourgeois public sphere" (Negt and Kluge [1972b] 1993, 95). Such unifying experiences were, in turn, to "offer forms of solidarity and reciprocity . . . grounded in a real collective experience of marginalization and expropriation" (Hansen 1993, 207), instead of shared abstract principles.

Finally, Negt and Kluge offered an alternative urban spatial ideal, one based on principles of familiarity instead of anonymity, and integration rather than separation. In making everyday life the foundation of political life, their argument implied that neighborliness and proximity were desirable forms of urban life. Bahrdt had defined *urbanity* in contradistinction to "the neighborhood" (Bahrdt 1952), and Habermasian planners saw high-rises, functionalist planning, and large public parks as facilitating bourgeois politics by sustaining the "tension" between the public and private spheres and allowing people to come together as strangers in public. Negt and Kluge helped the Eisenheim movement argue the opposite—that the colonies' small-scale, mixed-use character created "truly urban settlement units" of a completely different kind (Boström and Günter 1976, 92). In celebrating Eisenheim's "specific communication structure," which deviated both from "ideal-typical metropolitan standards" (Günter 1980, 15) and from the actual "metropolitan communication patterns" in the central city of Oberhausen (Von Merveldt 1971 quoted in Günter 1980, 15), their urban vision (if not their politics) was more in line with that of American proponents of small-scale urban neighborhoods such as Jane Jacobs and Herbert Gans.

## THE COLONIES AS A PROLETARIAN PUBLIC SPHERE

Like the *Revierparks* brochure, the movement's published materials are a self-conscious, self-made, and public relations–oriented guide to Eisenheim as a social and political space. Rather than creating a sort of user's manual for enacting a bourgeois public, their goal was to "make the streets speak": to explain how the "at first glance inconspicuous" street corners, walls, trees, and bridges add up to create a stage that supports the "baroque world theater" that is Eisenheim (Günter and Günter 1999, 11). In so doing, they also provide an account of how the Ruhr's new intellectual elite—its university professors and students—understood the colonies as helping to produce this proletarian, counterpublic sphere. While the *Revierparks* brochure used images of the colonies to represent insular, homogeneous, and closed life under National Socialism, the Eisenheim movement presented the colonies as spaces uniquely valuable for the construction of community.

Though Eisenheim was a housing movement, arguments for its preservation identified nature as a key source of its social and political goods. As depicted in the *Die Zeit* article mentioned above, it was the colony's green spaces and animals and the interactions and relationships that those conditions created that made the housing socially and politically valuable in the eyes of the movement. Gardens and animals are a central visual trope. The image reproduced in figure 4.2 is typical. On the left, a high-rise rental contract forbidding animals is shown, the handwritten addition to the contract helpfully specifying "No house pets!" By contrast, the caption that accompanies the image on the right says: "There are 399 animals here, and not only that." On the next page, the image on the left shows that it is "Forbidden to walk on the grass!" in a high-rise, while in the colony, the image on the right indicates, "the gardens are used by young and old." The book repeats this message throughout.

According to the movement, these spaces and their forms of working-class leisure and association did not represent some "dying mode of agricultural self-sufficiency," as they had for proponents of both garden cities and *Revierparks*; instead, they were sources of a new form of "leisure value" (Projektgruppe Eisenheim 1973, 122). That value was producing proletarian consciousness. A page from *Leben in Eisenheim* features a photograph of four men, leaning on a fence, standing in a garden, deep in conversation. The caption reads: "Conversations. They start with the potatoes and pigeons. And almost always end in politics" (Günter 1980, 128). How were potatoes and pigeons understood to produce proletarian politics? Their argument was that the density and types

FIGURE 4.2. Eisenheim's nature declared good. (From Projektgruppe Eisenheim 1973, 110–13.)

of interactions that animals and vegetables demand, and that green semipublic spaces allow, produce and maintain a community of socially skilled, solidary, imaginative adults. The movement's materials illustrate this point in a variety of ways.

## Green Rooms

First, the colonies offered a natural setting for the half-public sphere in the form of "green rooms." As the high-rise was defined by its functional and spatial separation from work and leisure, the colonies' governing principle was integration. This was achieved in design terms by the colonies' many *Zwischenbereiche* (in-between areas; Günter 1980, 79) or *Nebenräume* (adjoining rooms). Figure 4.3 shows social interactions characteristic of these overwhelmingly green ancillary spaces, such as one's "own ground-level entrance," "garden as kitchen," "conversation over the garden fence," and so on.

Green rooms turned Negt and Kluge's political vision into a planning principle. These spaces' soft edges, flowers, shade, and grassy lawns were the main points of social contact in the colony, and the Eisenheim movement described the majority of the social life of the settlement taking place in these spaces around, adjacent to, and outside the private space of the home. In addition to the photographic typology, another of the movement's books about Eisenheim provides a map and "communications analysis" of these spaces' use and a discussion of patterns of social interaction (at the front door, on the stoop,

Proletarian Counterpublics 123

at the corner of the house, from the window to the street, in the garden). The communications analysis offers a historical snapshot of how these spaces created the conditions for a half-public sphere, noting: "Most of the people linger on Werra Street, which forms a kind of courtyard of the settlement: a space that feels fairly intimate. The spaces between and behind the houses are used much more often than the streets. These half-public zones have an extraordinary significance: they are intimate like a courtyard, allow for a quick transition between interior and exterior spaces, and are therefore also used frequently. (However, multiple [types of] 'thresholds' [could potentially] lie between high-rise apartments and open space.) These spaces serve a wide variety of functions and are therefore irreplaceable" (Projektgruppe Eisenheim 1973, 82–83, 65, 83).

These "half-public zones" promised to transcend public/private distinctions of the sort that troubled Negt and Kluge and, interactionally speaking, accomplished objectives exactly the opposite of those of functionalist planning. They produced not tension and anonymity—as in *Revierparks*, between the formal public and the wholly private spheres—but spontaneity, informality, and familiarity. A frequently reproduced photograph is the image of a man and a woman having a conversation through a street-level window in the

FIGURE 4.3. Green rooms and other "in-between" areas. (From Günter and Günter 1999, 98–99.)

upper-left-hand corner of figure 4.3. The image is almost identical to the photograph of colony life in the *Revierparks* brochure (see chapter 3) but is offered with a very different interpretation. As the cover image of a photographic essay about Eisenheim's social qualities (Günter, Günter, and Henny 1982), it serves not as a representation of insularity but as a positive depiction of how a specific spatial form—a first-floor window low enough and large enough to hang out of comfortably—produces a particular type of social contact, that between public and private space. When reproduced in another context, it is captioned: "The incentive to frequently take up contact across interior and exterior spaces" (Günter 1980, 72).

The Eisenheim movement explicitly connected this spatial morphology and the interactions it facilitated to a Negt and Kluge–like political ideal: "All areas of open space in Eisenheim have a similarly differentiated relationship between the public and the private sphere—there arises, in certain ways, a half-public sphere [*Halböffentlichkeit*]" (Boström and Günter 1976, 67). While Habermas's bourgeois subject's "social ties are exclusively freely chosen and can be broken off at any time" (Projektgruppe Eisenheim 1973, 138), with interactions "stylized and prepared like rare visits to the living room in the high-rise" (Boström and Günter 1976, 65), in Eisenheim one might lean out the window to offer fruit, ask a favor of a neighbor on their way to the store, or call children to dinner. The movement characterizes interactions in this space as spontaneous rather than planned (Günter 1980, 102), informal rather than formal, and based on familiarity rather than anonymity. Debate is possible, not because of a public mask or mutual agreements among strangers, but because of an emergent set of linguistic and interactive norms that build up over time through these relatively informal interactions and the shared conventions of Eisenheim's *arbeiterspezifischen* (worker-specific) discourse (see Günter 1980, 100–102).

## Nature Makes Neighbors

If the colonies' green rooms are the setting for a counterpublic sphere, it is by interacting in, over, and across their flora and fauna that individuals become neighbors. In figure 4.4, a man proudly holds a rabbit aloft on the left, and residents young and old engage in "neighborly relations" (*Nachbarschaftsverhältnis*) on the right. The book from which the image is taken later declares: "People who keep a lot of rabbits are supported by their neighbors" (Projektgruppe Eisenheim 1973, 94–95, 123).

What is the connection between nature and neighborliness? The demands and the products of nature—the care it requires and the fruits it bears—organize

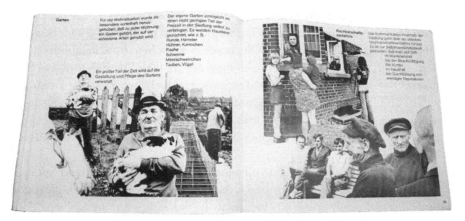

FIGURE 4.4. Nature makes neighbors. (From Projektgruppe Eisenheim 1973, 94–95.)

leisure time and increase social interactions. Rabbits create opportunities for neighborliness by attracting children, by having litters that need homes, and by requiring that a good amount of leisure time is spent out of doors, in proximity to the home. The garden creates opportunities for cooperation and sharing by producing more flowers than one person can pick or more vegetables than one nuclear family needs.

The Eisenheim movement also argued that interactions are not only quantitatively greater in a place like Eisenheim but also qualitatively different—marked by care and reciprocity. Affiliation and public concern were not motivated by prior, shared abstract normative beliefs, as Habermas imagined. Instead, "communal forms of association grow" from digging a plot of land, designing a garden shed, throwing a garden party, or exchanging fruit and flowers. "Animals" are, according to the movement, "toys, objects of curiosity, and communication bridges" that "offer learning opportunities for the development of caring relationships." (The text uses the word *Sorgeverhalten*, which means "custodial" relationships.) "The communication within the settlement goes beyond an ordinary neighborhood relationship. It is taken for granted that everyone helps: in case of illness, with supervision of children, in the household, when carrying out necessary repairs." Examples in the texts again attribute this reciprocity to earlier exchanges of nature. A resident, quoted discussing how he distributes lilies from his garden among his neighbors, goes on to describe a relationship between his sharing his lilies and his neighbors caring for his children when his wife is in the hospital: "I can go to my neighbor and say, 'Mrs. Selle, or Mrs. Adamschack, would you please—my wife is in the hospital—would you please look after the children until I come home from work?' That would be 100 percent accepted. But go into a high-rise and say that. . . . One never receives help from neighbors in a high-rise" (Projektgruppe Eisenheim 1973, 66, 76, 138). Sharing produces familiarity, which produces feelings of reciprocity, which turn into acts of mutual aid. In Eisenheim, "it is taken for granted that everyone helps" because, having dug a piece of land or exchanged fruit, residents come to know each other and develop trust and rapport—people who exchange flowers also help each other in times of need.

Beyond these individual, one-off encounters, "solidarity over work is created in the garden," as are organized forms of association. While, in a high-rise, "practically the only possibility for activity is [tinkering with] the car," gardening, construction, handwork, and animal husbandry are all popular hobbies in Eisenheim (Projektgruppe Eisenheim 1973, 123, 118–19). In part, these hobbies were products of constraints in a population that did not have the resources to buy records or magazines or frequent restaurants or movie theaters. But the care that gardens and animals require also meant that residents ended up spending the majority of their leisure time in the colony, together. One would not typically build a new garden shed or prepare a patch of ground for planting alone, while hobbyist pigeon keepers remark that they cannot go on vacation owing to their commitment to the birds' daily care (Soeffner 1997). The Günters describe the "associational life" that grew up around clubs devoted to pigeon keeping, gardening, and other hobbies as "companionable life" (Günter and Günter 1999, 143); other scholars of the Ruhr have argued that the clubs did not just "mak[e] miners into mates" but were a "decisive [pre]condition for [politics]" (Soeffner 1997, 114).

## Play as Emancipatory

The Eisenheim movement shared with mainstream planners and scholars of the day the worry that mindless, consumption-oriented free time was a threat to democratic politics, in that it dulled and pacified potential public citizens. Like *Revierparks* planners, and echoing the pastoral Faust of chapter 3, movement materials argued for uses of leisure time that were liberating and politically generative—and which were possible in the colonies but not in high-rises. In contrast to the "imagination-killing" high-rise, where children's "emancipatory dispositions [*Emanzipationsansätze*] are blocked" (Günter 1980, 173), it was argued that "all of Eisenheim is a playground" (Projektgruppe Eisenheim 1973, 122). Play, according to the campaign's imagery and materials, is a mode of "leisure education": the social learning time that liberates creative impulses, socializes children in the colonies' values, and produces socially skilled adults (Projektgruppe Eisenheim 1973, 122).

By requiring adults to spend the majority of their leisure time at home, gardens and animals foster relationships that transfer social skills and values to the next generation. *Rettet Eisenheim* describes the colony's "tinkering retirees and workers" as informal "leisure-time educators." "Leisure educator" (*Freizeitpädagogen*), they tell us, has become a profession in recent years, another task outsourced from the traditional middle-class home. But it is expensive to hire people to do this kind of work, and, in "newly developed areas" such as postwar high-rise communities, educators have a hard time finding children to educate at all. In Eisenheim, adults—often retired—engaged in leisure activities at home socialize and develop the skills of the young. Accompanying another arrangement of photographs celebrating house pets, one page explains: "Children who are able to move freely also move freely later in life. Not like in the high-rise, in a straitjacket. . . . The garden is a playground for our children. The area is ideal for children" (Projektgruppe Eisenheim 1973, 122).

In other words, children who get to engage in play become "free," expressive, confident adults. In *Leben in Eisenheim*, Janne Günter devotes an entire chapter to the colonies' role in the socialization of children, outlining the social skills that Eisenheim's supervised, intergenerational play creates. At home, and with neighbors, children have class-specific sociocultural experience, learn independence and try new things, negotiate group behavior and territory, and develop communication skills and emotional and social competence (Günter 1980, 162–203). Empathy and solidarity, productive conflict management, and self-assurance are opposed to the social consequences of living in a highrise, where the "anonymous and sterile neighborhood" leads to "criminality," "egocentrism," and "social crippling." She concludes with a quote from Oskar Negt, noting that "the foundation of any learning—even cognitive—is 'the development of emotional and social substructure'" (Negt 1978 quoted in Günter 1980, 204). Adult play in the form of hobbies provides analogous educational and creative opportunities. Adults in gardens repair and produce, help, lend, converse, and visit (Günter 1980, 116–17, 122, 125, 126, 132). These interactions "promote openness to each other and thereby trust—which is the basis of all social relations and all conscious community action. They promote complex exchange and thereby a variety of social-learning opportunities. They promote . . . complex thinking . . . understanding of human nature and thereby skills in dealing with people" (Boström and Günter 1976, 65). In sum: "In building a fence, man is his own architect" (Günter 1980, 118). For Negt and

Kluge, as well as for members of the Eisenheim movement, play of this kind is important because it develops the imagination. They interpret imagination or fantasy not as an escapist symptom of alienation but as an emancipatory act through which to explore alternative worlds and ways of being.

The kind of politics this all produces is, not a "formal-democratic process" as in the public sphere of the *Revierparks*, but an "informal discussion, on the street, in front of a window, in the garden" (Günter 1980, 69). And the participant is not an abstracted, rational political subject but one in touch with his or her own interests, class location, history, and creative capacities. Though the Eisenheim movement only rarely addressed the Ruhr's changing demographics directly, the influx of Turkish guest workers would have affected the colonies' residential composition. The argument was simply that, regardless of colony residents' background, familiarity, not anonymity, was the foundation for democratic politics. From informal conversations, forms of reciprocity, and interactive relationships, colony residents become "whole human beings" who are "trained, in cases of social conflict, to bring about prudent solutions because they have more understanding and practice in resolving conflicts" (Boström and Günter 1976, 65), and whose political subjectivity is a product of personal history rather than abstract convictions. The colony's "dense network of relationships" does not serve to differentiate or demarcate (*abgrenzen*) it from the world. On the contrary: "In the workers' public sphere trust in other people is developed." This trust is "the foundation of solidarity" for use in the broader public (Günter 1977 quoted in Günter 1980, 157).

## DISCOVERING WORKING-CLASS NATURE

Of course, the New Left did not *create* these forms of working-class culture. The Ruhr's working class had maintained its engagements with animals and green spaces—the number of allotment gardens (*Schrebergärten*, as they are known in Germany) and animal-keeping associations even exploded in number in the Ruhr in the first half of the twentieth century. But these types of engagement with the landscape had also been deliberately marginalized early in the century, and devalued by the dominant, bourgeois forms of politics and the postwar planning paradigm, with its emphasis on modern, comfortable, convenient living. The result was not only that the colonies were unpopular but also that their green spaces and agricultural practices did not actually register as nature in the eyes of the middle class. Neither colonies nor allotment gardens were mentioned when officials surveyed the existing network of green spaces in the Ruhr in the postwar years. They do not appear in the Ruhr's regional planning documents and are not depicted in the *Revierparks* brochure.

Of all the conditions that made the rediscovery of the colonies possible, the essential local factor was this material continuity, in the sense of the ongoing physical existence of these spaces and their importance to the Ruhr's working class. Since the nineteenth century, nature had remained a repository for working-class culture in the Ruhr in some very real ways—even if these were invisible to elites. As part 1 showed, in the nineteenth century, the colonies' gardens and animals had provided direct goods that served subsistence purposes, but, in the garden city era, the industrial barons rejected these forms of engagement with nature in favor of more urbane, cultured ideals for their workers. Drawing on urbanized nature, they transformed nature into bourgeois leisure, enforced through new rules for very similar spaces and activities, at the beginning of the twentieth century. The colonies and their agricultural associations were dismissed for a second time in the postwar era, when the end of wartime food restrictions, new consumer goods, and new democratic ambitions all suggested that the colonies should be replaced, and local planners offered new, liberal-bourgeois alternatives through high-rise housing and regional parks.

But, of course, the colonies had never disappeared, and the Ruhr's working class never stopped using or living in these spaces. In spite of the industrial barons' efforts at embourgeoisement, gardening, animal husbandry, and subsistence agriculture actually increased dramatically in the Ruhr in the first half of the twentieth century. Numerous small-animal and bird-breeding clubs were formed at the end of the century (Abrams 1992, 96), right around the time Margarethenhöhe was being built. In 1912, there were 39 organized small-animal-keeping associations in the region, and, in 1913, the League of Small Animal Breeders in North Rhine–Westphalia counted 468 such associations with 23,000 individual members (Abrams 1992, 122 n. 32). After the first law restricting labor to an eight-hour workday was passed in 1919 (kicking off the reduction in working hours that would reach peak leisure in the 1960s), the men, who had ceded work in gardens to women and children after the turn of the century, returned to them, raising rabbits and racing pigeons in addition to growing vegetables and keeping goats, pigs, and chickens (Steinborn 2010, 55). The allotment garden movement also exploded in the Ruhr between 1925 and 1933, with more than two million new gardens founded in these years (Steinborn 2010).

This, of course, was urbanized nature too, albeit in a proletarian rather than a bourgeois form. Urbanization transformed the symbolic meaning of nature for the working class just as it did for Ruhr elites. Just as for Krupp, gardens and animals lost their imaginative appeal as markers of an agricultural past and their economic function as a wage supplement at the beginning of the twentieth century. Just as for Krupp, gardens and animals came to serve social and symbolic rather than subsistence purposes—in small-animal-keeping associations and allotment gardens nature became an indirect, aspirational, and universal leisure good. But, unlike for Krupp, who had the means to build garden cities to realize a new urban ideal, the Ruhr's working class continued to associate with the same green spaces, gardens, and animals as they took on new meanings. In Durkheimian terms, these things became totems of the Ruhr's coal miners and steelworkers, reflecting aspects of the community, its ways of life, and its aspirations (Jerolmack 2013; Soeffner 1997).

Changing discourses surrounding the colonies' gardens and animals reveal these new symbolic meanings. In the nineteenth century, goats kept for milk had been the "miner's cow"; by the middle of the twentieth century, pigeons—kept for pleasure rather than sustenance—had become the "little man's racehorse" (Steinborn 2010, 56). Nature as leisure served a "compensation function," making up for long days underground (Günter 1980, 139). As one miner put it: "If one spends one's life among rough, dead stone, one longs for contrast, for soft 'material,' for light creatures, a rabbit or a pigeon" (Hoffmann 1982 quoted in Soeffner 1997, 107). Gardens also became leisure outlets rather than economic supplements. In the nineteenth century, industrial laborers leased nearby fields to grow food after the waged working day ended. By the middle of the twentieth century, colony and allotment gardens became places for tending roses and flowers, barbecuing, and socializing in the evenings and on weekends. In Günter's own interpretation, workers maintained a "strong bond to agricultural work" even as the "economic function [of the garden] shrank" (Günter 1980, 137), keeping pigeons and gardens "for the joy of the act" (Steinborn 2010, 55), and shifting focus from the "material worth of the harvest" to "treasuring" the garden as a "holiday island," a place to which to escape, a retreat.

And, of course, the more the subsistence role of colonies' gardens and animals declined, the more they became a primary means for the organization of social life. Wartime food shortages contributed to the expansion of colony and allotment gardens in the first half of the twentieth century. Such gardens were at that time essential for provisioning. They also served as important retreats during the French occupation of the Ruhr after World War I and again under National Socialism (Lüdtke 1995b). And, long before scholars and policy makers became conscious of leisure as an area of life to be managed, pigeons and gardens provided accessible, inexpensive opportunities for socialization and recreation. This is all to say that the Eisenheim movement's depiction of nature in the colonies was based on real observations about the way it had organized social life for the working class.

## SOCIAL REPRODUCTION AS THE NEW LOCUS OF POLITICS

A final, critical factor contributing to the rediscovery and reinvention of the colonies was deindustrialization, which, in shuttering the coal mines and steel factories, created an opportunity to reclaim formerly oppressive spaces for a Left politics. Factory closures prompted a shift in the center of political life from waged labor to everyday life, from political parties to leisure associations, and from production to the sphere of social reproduction. In spite of their different politics, the two versions of the public sphere depicted in materials surrounding these projects can be seen as two different responses to the pattern of inequality produced by a declining industrial economy in a region dominated by industrial production: massive unemployment among the Ruhr's traditional working class. Both are also political visions defined by leisure rather than work. While *Revierparks* planners deliberately bracketed the traditional foundation of politics in the Ruhr—class, labor, social relationships, and the workplace—by creating anonymous public green spaces, the Eisenheim movement found in the colonies a new source of class consciousness. As the industrial workplace disappeared, these inherently classed spaces began to look like places where class consciousness still existed and could be cultivated in the form of a distinctively working-class "culture."

Factory closures eroded the traditional basis of organized politics in the Ruhr. If the industrial workplace was no longer the center of economic or political life, it could no longer be the foundation of class-based organized politics. And, indeed, this was reflected in growing disappointment with the SPD in the 1960s and 1970s. As the traditional workers' party, the SDP had been the center of the Ruhr's political culture for a century and had also always contained both radical Marxist and reformist factions (Hancock 1978, 6). But there was a huge rise in West German support for the SPD between 1952 and 1972—in the Ruhr, the increase was from under 40 percent to around 60 percent of votes in federal, state, and local elections (Tenfelde and Urban 2010, 939)—that culminated in the 1969 election of Willy Brandt as the first Social Democratic chancellor. In this new role, the SDP confronted the challenges of economic restructuring by focusing on, not class or ideology, but the pursuit of "respectability" through policies of incremental, "pragmatic reform" (Smith 1976, 388). For example, it continued the previous Christian Democratic administration's support of "libertarian economic principles" and the free market despite its "earlier theoretical advocacy of partial socialization" (Hancock 1978, 8).*

Factory closures also changed the economy and management of the colonies themselves. Company housing transformed as the coal and steel industries collapsed. The practice of tying rental contracts to employment contracts had remained essentially unchanged since the 1850s, but providing housing became first a burden and then an absurdity as the workforce shrank. Employment at the former "Good Hope" coal mine and ironworks, for which Eisenheim had been built, dropped from thirteen thousand to thirty-five hundred between 1968 and 1987 (Kruse and Lichte 1991, 210–11), which meant that company dwellings housed fewer and fewer employees and any individual's tenure at a given company was tenuous (Nelles and Oppermann 1979, 51). When the Ruhr's remaining coal companies consolidated to form the Ruhrkohle AG (RAG) in 1968, RAG also took over management responsibility for the housing (Nelles and Oppermann 1979, 51). So, when in 1968 ATH bought the steelworks and the associated company property, including the land on which Eisenheim stood (Kruse and Lichte 1991, 210–11), the company proposed replacing the settlement with high-rises because it was receiving no tangible benefits from what was valuable land (Nelles and Oppermann 1979, 51), while RAG found maintaining the housing—by now in need of extensive repairs— increasingly expensive and a headache. By this time, the infrastructure and amenities in the colonies that had once been provided by the industrial barons had already been privatized or made public. In Essen, the grocery store chain

---

* These policies frustrated and alienated the party's more radical members. The most famous example was the local policy of *Mitbestimmung*, or "codetermination." The Ruhr coped with structural change within the "democratization" paradigm—by creating a "fabric of cooperation" at the regional and factory level—"despite diverging interests" (Goch 2002a, 93–94). *Mitbestimmung*, a policy introduced at the end of World War II, treated *participation* as a keyword for democracy by establishing worker representation at higher levels of management, including placing workers on companies' boards of directors. When introduced, codetermination was "a radical departure from the historical norm of factory authoritarianism characteristic of industrial societies" in that it introduced "parity representation between workers and shareholders" in the coal and steel industries (Hancock 1978, 17), but Left critics and radical socialists saw it as little more than a social palliative. By 1978, American scholars agreed that "private economic interests will continue to exercise decisive influence" (Hancock 1978, 18) even within such structures, and codetermination "emerged as one more legal formula rather than representing a breakthrough for democratic participation" (Smith 1976, 389).

ALDI took over Margarethenhöhe's *Konsum*, the library was liquidated and integrated with the city's public system, responsibility for health insurance was turned over to public insurers, the training workshop was turned into a vocational school, and the hospital was taken over by the state (Herbert 1983, 269).

As a result, the colonies—which had once represented a feudal paternalism that chained residents to their companies—could be imbued with new imaginative content. In the absence of an employer, it became possible to view company housing as a desirable and emancipatory space. The *Revierparks* deliberately demarcated a leisure "oasis" as the time and place for enacting democratic politics precisely because, by being separate from the home and the workplace, they removed consideration for social and economic difference from the sphere of public discourse. But, for Günter and the Eisenheim movement, the fundamental problem with the dominant bourgeois paradigm for urban development and democratic politics was its destruction of the "collaborative-solidary [*gemeinschaftlich-solidarischen*] social structure of the worker" and of workers' historical and political consciousness. The campaign argued that the bourgeois political ideal of "anonymity instead of solidarity" was not only a false promise of upward mobility but also a vision that would actually exacerbate inequality and deradicalize the working class by encouraging desires and lifestyles that would produce complacent, isolated consumers rather than active, engaged citizens (Projektgruppe Eisenheim 1973, 139, 138, 57). As an alternative, the movement turned the historical location of working-class domestic life into a site for the production of proletarian consciousness.

This new location for politics extended to leisure-oriented theories of solidarity. Work is often understood as the source of organized political action in the industrial era, such as the attribution of coal miners' propensity to strike to (among other factors) the fact that miners spend many long hours together underground. In the context of deindustrialization, as work disappeared, an "advantage [of the colony] to be highlighted" was, the movement argued, the fact that "a garden belongs to each unit," which made it possible to spend "a significant proportion of leisure time in the settlement itself," thus producing the same feelings of solidarity and mutual accountability (Projektgruppe Eisenheim 1973, 94–95). As the union had been to the workplace, the *Verein*, or "club," became the colonies' paradigmatic organizational form. The hobbies and associational life organized around these special interest groups became increasingly important for politics as mines and factories closed and disappointment with the SPD grew (Günter and Günter 1999, 143, 145).

By reinventing semipublic leisure spaces as political spaces, the Eisenheim movement turned these forms of daily life into the foundation for class-based politics. As one contemporary put it, while, "in the living world of the twentieth century," "the model certainly cannot be completely transported," the example of the colonies could be used to "develop evaluation criteria for workers' housing: for living situations where not anonymity but identification with the neighborhood, and not isolation but solidarity, determine quality of life" (Bösel 1974, 128).

## GREENING AND GOOD INTENTIONS

Though it is not obvious that old green forms could be reimbued with new and radically different content, this chapter verifies that such things are possible and outlines the conditions under which this occurred at one historical conjuncture. The reinvention of Eisenheim's nature and politics is just the kind of creative work for which Negt and Kluge—and Castoriadis—were hoping in the wake of the 1968 movements: that historically ideological or oppressive forms—whether company housing or mass media—could be sources of radically new forms of political consciousness. Nevertheless, Eisenheim was not turned from obsolescence into a "monument of proletarian culture" (Günter, Günter, and Henny 1982, 25) out of the blue, from pure imagination. A confluence of international political movements, local material transformations, and cultural trends made it possible to recast an environment of power and domination as an emancipatory one, interuppting almost a century of imaginative devaluation and physical decline.

In spite of the substantive differences in their political visions, as greening movements both the *Revierparks* and Eisenheim projects were outcomes of the same economic and political moment, and both responded to deindustrialization by turning to leisure and social reproduction as a space for democratic politics. Both were driven by good intentions—the '45ers' and then the '68ers' visions of democracy, respectively. And both were paternalistic, reproducing the same power dynamics between greening protagonists and receiving audiences even outside of a work context and in spite of their political commitments.

It is notable that the Eisenheim movement had a paternalistic aspect in spite of its revolutionary politics and cross-class membership. This was a product of the fact that the field in which it operated was that of the bourgeois public sphere. According to a survey completed by Günter's students, at the time of the movement, Eisenheim's residents were still the traditional working class: 25 percent miners, 25 percent ironworkers, 30 percent retired miners and ironworkers, and 20 percent railroad workers and mechanics. The great majority of them had lived there for over ten years, and only four of the married women were employed (Projektgruppe Eisenheim 1973, 64). And the movements to save the settlements were genuinely cross-class. When a workers' initiative organized to save the colony Rheinpreußen in Duisburg in 1977 went so far as to go on a hunger strike, participants were, "not students, but housewives and retirees from the settlement, as well as other population groups that are normally hardly prone to such spectacular actions" (Nelles and Oppermann 1979, 52).

But the Eisenheim movement found a high-status chronicler in Günter, and he and his university collaborators—his wife, students, and colleagues— were crucial to its success. They designated themselves "author-advisers" and "secretaries" of the movement (Boström and Günter 1976, n.p.), documenting the life of the colonies, producing books, pamphlets, and exhibitions, and organizing radio and television appearances to promote the colonies' importance to a local and international audience. Though they described this as a simple division of labor—skilled in writing and speaking, they were the ones who produced the records, the press statements, the radio interviews, and the newsletters—they were not so much the movement's secretaries as they were its translators. In spite of the fact that they had no official authority over the communities they represented, and in spite of their explicit efforts to relinquish what authority they did possess, their participation was key to granting the movement broader legitimacy and ensuring its success.

Negt and Kluge have, like contemporary critics of Habermas, argued that there are "language barriers" to entering the bourgeois public sphere, given the specific "economy of speech" for each realm (Negt and Kluge [1972b] 1993, 45). That these actors described their role as secretaries downplayed the degree to which the movement's success was due to their fluency in communicating the value of the settlements to the elected officials and conservation professionals who were the ultimate arbiters of the Eisenheim dispute, and their ability to pull strings, mobilize social networks, and make use of social capital. Günter, for instance, who had worked at the *Landeskonservator*, or department of historic preservation, made phone calls to get that office on the side of the movement and mobilized federal legislation creatively in support of the cause (Nelles and Oppermann 1979, 60). In spite of the fact that the Eisenheim movement was dreaming of a world beyond existing social relationships, the ultimate goal—to convince the public sector to step in and preserve the colonies—was still dependent on protagonists' ability to speak the language of the bourgeois public sphere. This moved Günter into a paternalistic relationship to Eisenheim's working-class residents, as he was called to represent their interests and speak on their behalf, in spite of his own intentions.

Examining the movement also adds to our understanding of the social logics of urban greening and urbanized nature that will continue to be explored in part 3, specifically regarding their relationship to power and inequality. In showcasing several axes of variation in greening's politics and content—a top-down Left project, a reimagining and reinvention, and a glimpse of popular desires (even if communicated through elites)—this chapter documents a heterogeneity of greening practices that is much closer to what greening looks like in real life as an everyday practice and that is not visible in other chapters. That is because this book deliberately traces urbanized nature through large-scale, top-down projects, following protagonists with the resources and moral or legislated authority to green at the scale of the city—to imagine an ideal urban public, to have those visions be accepted as legitimate, and to have access to resources to carry them out. However, urbanized nature's availability for multiple visions in the same period and specifically for a radical, grassroots project underscores that this imaginary was available not just to elites and that greening practices are not inherently system affirmative. While practicalities of scale, cost, and influence contribute to the fact that the most visible greening projects tend to be those that use nature to shore up existing social arrangements, Margarethenhöhe's and the *Revierparks*' bourgeois bias and orientation to the status quo is a product of the conditions of their construction rather than any qualities of urbanized nature itself. In the same decades, urbanized nature made it possible to preserve a competing set of ideals through other natures elsewhere—in the colonies, where gardens and animals remained a foundation of everyday life in ways that were largely invisible to elite reformers.

In retrospect, the movement to save the settlements was also historically significant because it prefigured the contemporary interest in industrial heritage that so defines the Ruhr today and to which we will turn in part 3. Eisenheim's preservation could be seen as the beginning of the Ruhr's discovery of "the value of the material and architectural traditions of its industrial history" in the 1970s (Goch 2002a, 102). This "paradigm shift" was helped along by artists and writers as well as residents and concerned intellectuals (Mesecke 2010, 244) and was also part of the growing interest in representing the Ruhr's disappearing industrial working class. Local artists and literary figures were part of this movement, as were international ones such as the renowned photographers Bernd and Hilla Becher. Hilla Becher grew up in the Ruhr, and the couple's now-iconic images of water towers, blast furnaces, gas tanks, and industrial facades were some of the first to aestheticize the Ruhr's industrial architecture. The first major preservation success in the region was the designation of Dortmund's Zeche Zollern colliery as a monument of historic significance in 1969, three years after its closure. Zeche Zollern became the headquarters of the Westphalian Industrial Museum in 1981. Today it is not hard to see the brick building as beautiful, with its Art Nouveau stained-glass entrance and its cathedral-like design, both part of the celebration of industrial production as a "palace of labor" at the time of its construction. But, in the 1970s, few saw these as monuments worth saving. This growing recognition and celebration of industrial heritage would form the basis of a tourism- and consumption-based economy in the 1990s, to which we now turn.

# * 3 *

# *The Social Life of Urbanized Nature*

CHAPTER 5

# Producing Nature, Projecting Urban Futures

By 1989, after three decades of job loss and mine and factory closures, the decline of the Ruhr's industrial economy was mostly complete. It had left a landscape riddled with thousands of acres of industrial "wastelands": former extraction and production sites now empty, polluted, and unused. For the public, these unsafe and unsettling places were a constant reminder of the Ruhr's industrial past and uncertain future.

The year 1989 also marked the beginning of a third greening moment in the Ruhr. The Internationale Bauausstellung (IBA) Emscher Park was a decade-long, region-wide, public-sector-led effort to recycle these wastelands by turning 457 square kilometers of brownfield into a giant regional park. With IBA, we enter a well-known greening era, during which the phrase *urban greening* actually originated, and in which such brownfield redevelopments and green industrial aesthetics have become popular and familiar. The IBA project was one of the first of its kind and in many ways pioneered this model. Though it is hard to imagine today, when IBA was funded in 1989 (just before reunification, when the Ruhr's problems still overshadowed those of the former East Germany), the idea of postindustrial green space did not really exist yet, as either a recognizable aesthetic or a replicable economic development strategy. IBA's regional directors—experts in urban planning, landscape development, urban policy, and social science—envisioned and planned the massive project before the construction of New York's High Line, before Paris's Parc André Citroën, before the public celebration of urban agriculture in Detroit, and before the rise of urban sustainability as a popular policy paradigm.

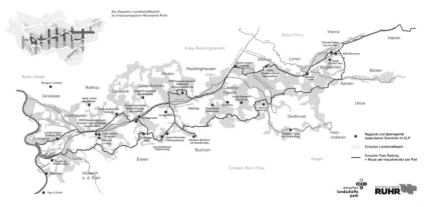

FIGURE 5.1. IBA's Emscher Landscape Park. (Courtesy of Emscher Landschaftspark and Regionalverband Ruhr, 2017.)

As conceived, IBA Emscher Park would connect seventeen cities in the poorest part of the Ruhr with bicycle and pedestrian paths, cultural heritage sites, public art, and ecological restoration projects. And, with the park as an anchoring concept, the project facilitated the €4.4 billion restoration of the "biologically dead" Emscher River (Girardet 2014, 147), converted industrial buildings into museums and cultural destinations, and organized or commissioned over a hundred projects oriented toward ecological restoration, urban development, housing, tourism, and recreation. Today, the park is about 43 miles long and 125 square miles, and it contains over one hundred museums, heritage sites, art installations, and monuments, including the Ruhr Museum, located in an old coal-washing plant, designed by Rem Koolhaas (fig. 5.1). "Essen for the Ruhr" was named a European Capital of Culture in 2010 in large part as a result of these investments, and the Emscher Park Master Plan was produced in 2010 to continue realizing IBA's vision.

The project aimed to reinvent the Ruhr's industrial past and provide a template for its economic future by advancing a new, positive image of the region as one whose history had uniquely shaped its culture and ecology, with the IBA as its primary urban development tool. IBAs, or International Building Exhibitions, are, as a friend likes to joke, neither international, nor buildings, nor exhibitions, but concentrated urban development projects that provide dedicated funding, staff, and legislative power to undertake planning interventions with an eye toward modeling "the city of the future" for a set period of time and with an agreed-on theme (Shay 2012, 19). They are part of a history of planning exhibitions in Germany that goes back to the beginning of the twentieth century. Perhaps the most famous IBA took place in West Berlin from 1984 to 1987, where IBA provided housing, day-care centers, schools, libraries, and other infrastructural improvements in order to repair the social fabric of Kreuzberg, a neighborhood ravaged by war and splintered by the Berlin Wall (Miller 1993). Led by the state of North Rhine–Westphalia (NRW), and carried out with regional, federal, and EU funds, IBA Emscher Park focused on the Ruhr's Emscher subdistrict, a band running across the most densely settled middle-northern part of the region, which, unlike the medieval market towns in the Hellweg zone farther south, lacked any traditional urban centers (Kilper and Wood 1995, 208), and whose two million residents were disproportionately poor and of migrant background (Reicher, Niemann, and Uttke 2011). A 1989 project memo described it as "the most densely built-up industrial landscape in Central Europe with the highest levels of environmental pollution and the most intensely carved-up open spaces" (Kilper and Wood 1995, 209).

Part 2 argued that nature is understood to be particularly useful for urban publicness as a result of its apparent universal accessibility and the universality of experience it provides: it is presumed to be beneficial to everyone and to everyone in the same way. Part 3 takes advantage of data available in the present to show how urbanized nature—and particularly this belief in nature's universal benefit—shapes the dynamics of greening projects' production and reception. Chapters 5 and 6 examine urbanized nature in the wild, so to speak, by looking at the social life of this social imaginary. Chapter 5 documents the production of IBA Emscher Park from the perspective of greening protagonists in order to understand how greening projects are actively constructed as universally beneficial nature. IBA's directors' descriptions of the behind-the-scenes work involved in creating a nature experience, supplemented by my observations and those of critics, reveal that greening protagonists are not naive about what the sociologist David Grazian (2015) has referred to as "nature-making" efforts. They can describe the significant imaginative work, and the work of image production and reception, in which they engage in order to get people to see and inhabit greening projects and urban spaces in the way they intend. But their ultimate goal is to provide an experience of unmediated access to the landscape. These accounts show how greening protagonists' awareness of the social organization of nature experience coexists with a subsequent discursive (and, ideally, experiential) erasure of those efforts, the step that is required for green landscapes to be viewed as nature.

## THE URBAN FUTURE AS A PROJECT

Interviews, more than historical data, make it possible to examine protagonists' self-perceptions of their greening acts and, therefore, to see imaginative work in action. They document what that work consists of, descriptively, and make concrete what it means to say that greening projects are imaginative projects. I observed this in a 2013 interview with Michael Schwarze-Rodrian, a landscape planner who was one of the conceivers and directors of IBA Emscher Park, in which I kept asking the wrong questions. Schwarze-Rodrian was, at the time of our interview, the representative of the RVR (the Regionalverband Ruhr, the contemporary name for the SVR, the regional planning organization founded by Robert Schmidt and discussed in chapters 2 and 3) to the European Union. I began with what I thought was a fairly straightforward question about the park: "How was the land secured?" To which he responded: "The Emscher Landscape Park is not something to own." Trying to wrap my mind around this—maybe it was some kind of guerrilla restoration project?—I asked: "So you just looked for brown fields and . . . ?" "No no no," he said. IBA Emscher Park's task was not to create a park where none existed. Instead:

> The urban reality is there. And our idea was, How can we qualify this reality? Sometimes the land belongs to an old mining company, sometimes it belongs to a farmer, sometimes it belongs to the city, sometimes it belongs to the electricity power supplier, sometimes it belongs to the Emschergenossenschaft [the Ruhr's regional water management body] as a river system, etc. And we had a holistic view from the very beginning. We said, Where is the corridor? And, within this corridor, we allow every kind of question in the sense of qualifying. In the sense of, What is going to make it better? We had no operating idea that we had to make it [the park]. We didn't have the idea that after we have discovered it we were directly responsible for making it. We had the idea that it was senseful [*sic*] to think, to work together to find people who shared this idea and who would like to invest, etc.

In other words, IBA Emscher Park was an imaginative project—literally. The park itself was a concept. From the beginning, IBA's project directors understood their goal as "qualifying" the existing urban reality, to get people to see the landscape differently or, in other words, to change the social imaginary of the region. I tried to probe what it meant that they did not really have to "make" the park, that the "urban reality" was already there, with a question about marketing, but Schwarze-Rodrian interjected: "We are not going into cinema. We are not going into TV. He [Karl Ganser, IBA's director] was not saying, 'I just have to change people's minds, and then we have a park.' We invested €1 billion of public money in the park. It was a great investment by the state of North Rhine–Westphalia that we said was necessary. Refurbishing the Emscher [River] cost €3.4 billion, and we're still paying."

Eventually I caught on. To say that the project had immaterial objectives did not mean that it was not also physically significant. As Schwarze-Rodrian was trying to impress on me, IBA Emscher Park was not cinema. The directors were well aware that "images alone can achieve little" (Sieverts [1997] 2003, 111), and the project involved massive investments in physical engineering, ecological remediation, and new construction. But these investments were all geared toward making it possible to tell a new story about the Ruhr and carried out with the goal of helping people learn to see the region in a new way. Schwarze-Rodrian's point—as Walter Siebel, an urban sociologist also on the IBA team put it elsewhere—was that "achievements . . . made visible via an 'exhibition' . . . are meant to consist of mainly non-visible outcomes such as changing the social, cultural, and mental structures of the region" (Siebel et al. quoted in Danielzyk and Wood 2004, 202). From the beginning, IBA's directors' goal was to catalyze an imaginative shift, or what Ganser described as "mental change" (Ganser 2010, 9); they understood this "to a large extent" as both a "cultural" and a "planning" task (Sieverts [1997] 2003, 81).

The sociologist Ann Mische (2009) has described the future as a "project"— something embarked on, set out on, imagined and "projected," if you will, into space and time through material and discursive practices. She argues for the "constitutive role that the future imaginary plays in reflective processes of critique, problem-solving, and social intervention" in the present (Mische 2014, 440). Future-oriented practice of this kind is just the sort of work that this book documents and that I have called *aspirational*. Urban greening projects are projections of ideal futures that shape social space and social action in order to create them; like each of the projects described in previous chapters, IBA Emscher Park was a project of creating an urban future. But, as my questions to Schwarze-Rodrian reflect, imaginative practice can be a hard thing to grasp. It is easier to understand actually creating a park, providing access where none existed—as Ruhr planners created *Revierparks* in the 1970s—as a means of changing reality. The purpose of IBA Emscher Park was not to make some objective nature newly available but to get people to interpret the existing landscape in a new way.

Mische focuses on verbal and visual representations rather than material form (Mische 2014, 442). We, however, are by now well prepared to explain why green was the chosen vocabulary for this project in the first place. As previous chapters have shown, in the preceding hundred years greening had become a *habit*: the idea of nature as a medium for improving the social took root in the minds of European and, later, American elites as they worked out what modern industrial cities should be and be like—and spread. As people became habituated to its use, greening became widely legible as an aspirational practice. It was the availability of this idea of nature that made it possible for a park to seem like a logical and creative way to reenvision the Ruhr. Schwarze-Rodrian himself—then a self-described "crazy young university guy with cuckoo ideas"—was the one who initially proposed the park concept in response to Karl Ganser's call for proposals. While Robert Schmidt had initially designed the Ruhr's seven regional greenbelts to "protect" the region's remaining open space and limit and shape existing settlements in the 1920s, half a century later, Schwarze-Rodrian told me, the concept was to "turn it around," instead making the greenbelts the spine of a regional landscape park that connected cities rather than separating them.

In drawing attention to the actual work that has to be done for such transformations to take place and for such ideas to stick, Mische encourages us to ask, How is such an idea produced and reproduced in the minds of the public? What did IBA have to do to get people to be able to see the region as a giant park? The answer, which this chapter explores, is that it requires a tremendous amount of work for imagined futures to be socially realized. As I struggled to wrap my mind around the real/not real, imaginative/material quality of IBA Emscher Park, Schwarze-Rodrian told me this story about the skepticism and resistance he encountered when Ganser commissioned him to undertake a year-long feasibility study:

> We were frustrated. . . . Whenever we talked to the cities, they said, "Are you cuckoo? A park in the middle of the Ruhr? The landscape is outside, the nature is outside, and you must be a green dreamer." Or more aggressive, "You are a Green, a political man, and we don't do things like that." The only thing that kept them from being unfriendly was the ticket saying that it was Karl Ganser, who was the head of urban development in the state, who asked us to do this job. And so they said, "OK, we can talk about it, but . . . we don't believe you, we don't believe in it, and tell Ganser we're not interested." So we came back with maps full of options and protocols full of [people saying] "not interested."

But, Schwarze-Rodrian reported, Ganser was unfazed, saying: "What a wonderful result! If you'd found the opposite, we couldn't do anything—if there were no sites, but just lovers of a park, nothing would happen—but you have found the sites. The people do not understand, but we can change that! We cannot change the existence or nonexistence of sites and potential."

Because the sites existed but the people did not yet understand the concept, IBA began a strategic redevelopment of the landscape to change people's perception of the region. There was a real, material basis. There had to *be* greenbelts, brownfields, a foundation of open space. There had to be room for a park. (Which of course there was, thanks to the region's polycentric structure and planning history.) But creating the park was not the bureaucratic problem of land acquisition and management that I had initially imagined. The challenging part was what those involved described as the "cultural" task of getting people to see the Ruhr in a new way. When Schwarze-Rodrian discovered that the park was "physically possible but imaginatively impossible," he meant that people (whether locals or outsiders) had not yet been taught to see the Ruhr as green.

As prior chapters of this book have demonstrated, though it is more common to think about providing access to nature in straightforwardly physical terms, all greening projects are also imaginative projects. The existence of materials created to communicate these visions to public audiences—of the sort used as data throughout this book—belies the fact that, even if protagonists greened believing that nature was transparently relatable and universally beneficial, they still had to do significant social work to get people to see and inhabit nature as they intended. In 1910, Margarethenhöhe's house rules reinforced the new ideals of citizenship communicated through the garden city; in the 1970s, both the *Revierparks* planners and the Eisenheim movement relied heavily on publications that projected future Ruhr cities and citizens onto the page for people.

This chapter shows what this cultural work consists of, drawing on greening protagonists' accounts of their own actions to provide a descriptive account of the actual labor involved in socializing people to these spaces and visions of the future. With the idea of a park as a heuristic for the larger project, IBA's goal was to help people narrate the Ruhr's landscape differently, to make strategic interventions in this field in a way that enabled people to see the region as beautiful and something to be proud of. The big question on the table, as Schwarze-Rodrian put it, was "how to make it."

## ECONOMIC GLOBALIZATION AND GREENING UNDER NEOLIBERALISM

Though IBA Emscher Park is a much more familiar example of greening as a contemporary, postindustrial phenomenon, equipped with knowledge of prior greening periods it becomes possible to tease out which characteristics are unique to this moment and which are more general attributes of greening as a social practice. Globally, the late 1980s and the 1990s marked another period of major urban restructuring, often referred to as *neoliberalization*. The term describes a package of regulatory responses to the crisis of the Keynesian welfare state and the end of the industrial economy beginning in the 1980s, as well as an "ideologically drenched" form of globalization, a type of national regime, uneven, global forms of market-oriented reform, and a free market ideology that became dominant during these decades (Brenner, Peck, and Theodore 2010, 183). This was also the era in which *urban greening* came into existence as a policy term, the greening of postindustrial landscapes became popular, and flagship urban-industrial greening projects such as New York City's High Line were constructed. Documentation of the unequal consequences of such projects has followed. From New York's High Line (Lindner and Rosa 2017) to its waterfront (Vormann 2015), from Paris (Newman 2015) to Bogotá (Galvis 2014), scholars have found that contemporary greening projects, often reliant on private funding, have tended to produce privatized, quasi-public green spaces that stratify consumer publics along class lines, raise property values and leave residents vulnerable to displacement, and frequently lack democratic accessibility in terms of acceptable users and forms of usership.

As a green response to urban-industrial economic and regulatory pressures, the IBA project in many ways fits this story. The Ruhr experienced the 1980s and 1990s in a manner similar to that in which cities such as Pittsburgh or Detroit in the United States did: through the globalization of industrial production and manufacturing. Economically, globalization killed what was left of the local industrial economy. Environmentally, it brought about what the *New York Times* described in 2007 as a "pollution transfer" from West to East. An article and an accompanying short video tracked the sale and relocation of the Phoenix steel plant and blast furnace from Dortmund, the largest city in the Ruhr (and the one in which I lived for a year), to China. It presents the disparate fates of the two towns as twin effects of the same process of global economic restructuring. A Thyssenkrupp representative describes the work of four hundred Chinese laborers to dismantle the factory piece by piece and then pack, transport, and ship 220,000 tons of material to China, where, rebuilt in Hangang, the factory "cover[s] four square miles and resemble[s] a working museum of the industrial age." There, it fuels economic growth, but it has also produced "cancer towns" where local cabbages are too polluted to sell and to which traveling outdoor film festivals refuse to come because the ash in the air makes projection impossible. Sounding much like accounts of the Ruhr in the 1950s, the article describes "shimmering yellow and raging red . . . flare stacks [that] burn off waste gasses and inflame the night sky," a "noisome, noxious, money-spinning, job creating leviathan," and people who suffer increased risk of lung cancer and cannot eat outside due to the coal dust in the air, protesting, "Don't darken our skies!" (Kahn and Landler 2007). The article and video suggest that Germany's higher production costs and emissions standards have become China's polluted cabbages and that Germany is on the winning side of this exchange, having traded its steel and smokestacks for light manufacturing and alternative energy.

But, for cities like Dortmund or Detroit, economic globalization also left massive physical, economic, and emotional holes in its wake. Coal mining and steel production are expansive and intensive land uses, and the end of the industrial economy meant that one thousand hectares in the Ruhr served as a daily reminder of loss (Dettmar 2005, 264). Dortmund alone had over two hundred hectares of such spaces, or, as the Phoenix plant's project website puts it in a more locally meaningful quantity, an area as large as three hundred soccer fields ("Tradition und Moderne" 2014), much of it heavily polluted and increasingly dangerous as industrial buildings began to decay. Thus, while the closure and sale of the plant did mean less soot in the air, less noise, and cleaner laundry, as any Ruhr resident will remind you it also meant the loss of the eight thousand jobs remaining from what had once been a workforce of forty thousand (Jonas 2008) and the final darkening of the "land of 1000 fires," referring to the blast furnaces that had once been a reassuring image of prosperity in the region (Danielzyk and Wood 2004).* In addition to unemployment, and long past the temporary strangeness of watching a factory being dismantled, former miners and factory workers also lived with daily, physical reminders of globalization. Most of the employees of the Phoenix plant had lived in the

---

* The actual relocation of the plant was really the proverbial final straw. Until shortly before its closure, the Phoenix plant had been owned by a company called Hoesch that in many ways was Dortmund's Krupp. Formed in 1966 through a series of mergers of smaller companies, at that time it consisted of four interconnected steel-production sites that employed forty thousand people (Jonas 2008). Krupp acquired Hoesch in 1992 as part of ongoing, industrywide consolidation, and, in 1997, Hoesch-Krupp and Thyssen (the Krupp of Duisburg) merged as well, forming Thyssenkrupp in a final effort to preserve any steel production in the Ruhr. Because Thyssenkrupp would focus its operations in Essen and Duisburg, where each company had originally been located, most of the eight thousand jobs cut in the merger came from the former Hoesch plant in Dortmund (Hudson and Swanton 2011, 12), and, in 2001, the company closed the Dortmund steelworks for good.

adjacent neighborhood of Hörde, and now, when they looked out their windows, instead of seeing one thousand fires they saw a wasteland.

These physical transformations required IBA's project directors to tackle landscape and land use in a big way. Though the topic of structural change was, by the time of my fieldwork in 2012, mundane for local residents, the changing landscape remained shocking—or at least interesting—to people, perhaps because these transformations tracked economic change. More than one interview ended crouched over a map or leaning over a computer screen watching people trace their fingers along the outlines of former coal mines, factory sites, or railroads. "This must be an old photo," they would say, pointing at a building or what looked like a construction site on Google Maps. "This is no longer here." Or, "This is a lake now." Or simply, "This doesn't look like that any more." In 1999, Walter Siebel observed the significance of these physical changes (and absences) as mirrors of economic and cultural change, writing of the region: "Now that the heavy industry which originally created and characterized the region is disappearing, a lot more than just the economic base is being lost. The heart of the community has also been destroyed, not just in physical terms, but in terms of its social, political and cultural core. The factory was at the heart of the community. It was a place of domination and exploitation, as well as the seat of power and an object of pride for the working community. All that is left today is an enormous, gaping hole." Therefore, in addition to economic restructuring, job training, and attracting new businesses—all the necessary economic tasks—the physical landscape and its meaning had to be dealt with as well. Siebel described this work of filling "empty space" with a "new, distinctive urban quality" as "one of the most difficult urban renewal tasks," one that inflamed planners' "horror of empty spaces" and the impulse to erase history and negative reminders of the past (Siebel 1999, 124–25, 128, 130).

Partly as a result of projects like IBA, the impulse to fill, hide, and erase industrial legacies is perhaps less often the first response than it was in the 1970s. In the past few decades, another formula has emerged: the transformation of industrial brownfields into postindustrial, income-generating green fields. Along with commodifying industrial sites as cultural heritage sites beginning in the 1970s and 1980s, city governments have learned to recycle former working waterfronts or polluted brownfields as sites of leisure, tourism, and consumption for new urban economies. In an environment of territorial competitiveness and "fast policy" transfer, as cities compete to attract workers and capital investments and develop new forms of economic growth (Brenner and Wachsmuth 2012; Peck and Theodore 2015), and as green was becoming a tool for urban economic redevelopment in the post-Fordist era (Greenberg 2015; Keil and Graham 2005), entrepreneurial urban governments dealing with an impoverished public sector and massive cuts in public funding and services scrambled to convert industrial manufacturing, logistics, and transportation sites into recreational green spaces for a new wage-earning class.

This is the context in which the great majority of scholarship on urban greening has been written. The hope is that such plans will kill two birds with one stone, replacing postindustrial wastelands with public recreational space and lost economic production value with consumption value. As economic spaces, greened industrial areas serve as valuable public resources and contribute to economic growth either directly, by generating income and jobs through on-site concessions, or indirectly, by attracting businesses, residents, and developers to an area with a good quality of life. As municipal funding cuts and cities' decreased capacity for investment in public goods have put increasing pressure on spaces like parks to be income generating and self-sustaining, they also, increasingly, earn the money for their own maintenance. New York City's Brooklyn Bridge Park, for instance—also a reinvention of a former industrial area created by economic globalization—earns the $16 million annual maintenance costs for its public leisure areas and sports fields through income generated by on-site LEED-certified luxury condominiums and commerce (Gould and Lewis 2016, 69). In the Ruhr, as we will see, private waterfront homes underwrote the cost of public waterfront recreation, just as, in New York, those who buy apartments in Brooklyn Bridge Park make it possible for the rest of us to buy ice cream there or take in the view.

In the context of a longitudinal comparison, neoliberalization—as an economic and regulatory moment—is useful for highlighting the IBA project's major points of contrast to greening in prior historical periods, especially its conception of urban publicness and the perceived relationship of the parks to consumption. Neoliberalism is also often associated with the dismantling of the postwar public sphere. As city budgets were cut and public spaces privatized, new economic constraints intersected with commitments to the public good to produce new understandings of urban publics as consumer publics (Madden 2010; Scammell 2000) and of citizenship as a practice of consumption (Baudrillard [1998] 2016; Zukin and Maguire 2004; see also Krinsky and Simonet 2017). While in some ways continuous with the shift to public consumption described in chapter 2, this was a very different model than that of the postwar era, in which West German planners feared the overgrowth of consumption-oriented leisure time and planned the *Revierparks* to help mitigate that by providing spaces that offered alternatives to consumption. But, in the 1990s and especially the early years of the twenty-first century, as chapter 6 will show, planners accepted the idea that parks could generate revenue and that users—whether international tourists or local weekenders—were also consumers of goods and services. This translated into a new understanding of park spaces, not as public goods that are freely provided, but as amenities made possible by other compromises, and that are public to the extent that all are able to consume according to their means.

Though IBA Emscher Park was part of this green economic development moment, in some respects it was an unusual project. The IBA's European context, ample funding, and public charge gave it a strong civic orientation. It was also self-consciously an "urban," rather than an "economic," development project (Danielzyk and Wood 2004), with its funding coming from the NRW's Ministry of Urban Development, Housing, and Transportation (Rehfeld 1995, 211–12) rather than Economic Development or Environmental or Spatial Planning. This gave it greater jurisdictional authority and made it less market oriented in its execution. While IBA Emscher Park had "soft" economic goals, it was carried out alongside other "hard" economic regionalization projects geared toward promoting economic diversification and investment in technology (Danielzyk 1992; Brenner 2000, 325). It explicitly focused its efforts on the poorest and most polluted part of the region and, like other IBAs, took affordable housing and community development as part of its mandate. And its leadership was described as having had an "intellectual character," which is perhaps a nice way of saying that the project was very conceptual and was run by academics (Adams and Pinch 2011, 192). Karl Ganser, IBA's "visionary" and "sulfurous" director (Provoost and Vanstiphout 2011, 264), was a geographer and urban planner whose team included social scientists such as Walter Siebel and Thomas Sieverts, an influential architect, planner, and theorist of urbanism. So, although its aesthetics are familiar, the project's significant public funding, broad administrative authority, and social democratic commitments all placed it somewhat outside the standard greening paradigm.

The objective of this book, however, is to explain greening as a social practice and to identify the durable social, political, and imaginative characteristics that can be obscured in contemporary discussions of green neoliberalism. When IBA Emscher Park is viewed in the context of prior greening periods, a common thread of underlying good intentions emerges. Like greening projects examined in earlier chapters, IBA was shaped by its political, economic, and regulatory context in the sense that the ideals to which it aspired and the forms in which those aspirations were expressed reflected the political economy and modes of regulation of the era (Steinmetz 2005; Sum and Jessop 2013), whether they affirmed dominant beliefs and social arrangements or, as in the case of Eisenheim, were a reaction against them. But, even as greening projects reflect the historical context in which they are embedded, contemporary interview and observational data bring protagonists' moral mandate and public ethic into focus: how they see greening as an improvement, as a public good, and how their activities are driven by good intentions and described as such, even in the face of the projects' material compromises and constraints. The remainder of this chapter examines what, thanks to urbanized nature, people understand themselves to be doing when they green. It finds that IBA Emscher Park's directors envisioned and offered the park as a public good, one understood to model a more active role for government, a higher expectation of public support, and a different vision of democratic public space than the majority of industrial nature projects of the period.

## REINVENTING THE RUHR

One thing the Ruhr's history of successive transformations of the landscape makes clear is how much of the imaginative work of greening is really *reinvention*. Though it is most often marginal spaces leftover from industry or war that we think of as wastelands today (Gandy 2013), in each period political and economic change creates opportunities for greening by devaluing certain types of sites and spaces, leaving them available for economic recapitalization, ecological remediation, and social revaluation. As Margarethenhöhe's architect helped reinvent agricultural farmland as a green leisure landscape in the nineteenth century, as planners treated former agricultural land as a blank slate on which to build regional parks in the 1960s, and as New Left activists reenvisioned company housing as emancipatory in the 1970s, in the 1990s IBA began to see fallow industrial land as recreational green space. The holes left by the globalization of industrial production and local economic decline became apparent wastelands to be reinvented as nature.

The particular physicality of industrial production did limit IBA's options for reinvention. Because it would have been unthinkably costly to demolish the massive industrial buildings and infrastructure, creating a blank slate was not really an option. "Pre-industrial clichés of picturesque and orderly woodlands" were "useless" as models for ecological restoration in the heavily mined and polluted landscape (Weilacher 2010, 88–89). And the window for repurposing was closing quickly. Fenced and gated, leaking gas, littered with heavy machinery, and shedding cracking brick and concrete, the sites were becoming dangerous, hard to secure, and expensive to rehabilitate.

In response, IBA came up with an approach it called *recycling* or sometimes "change without growth" (Wachten 1996), by which it meant working with what is available, with attention to history, cognizant of a shrinking economy. Within these constraints, Ganser and his team of charismatic visionaries came up with a narrative that also happened to invert two of the major binaries of modernity: the relationship of city and countryside and that of nature and industry. In 1989, the proposal to understand the Ruhr's industrial past not as something that had *destroyed* its nature and culture—by extracting coal, polluting its waters, and exploiting its labor—but instead as the source of new, unique forms of culture and ecology was quite a radical one. IBA's directors proposed a development centered around an industrial-environmental aesthetic and urban-environmental consciousness, encapsulated by the German terms *Zwischenstadt* (in-between city) and *Industrienatur* (industrial nature). These concepts offered locals a way to narrate personal and family history as a source of pride, rather than shame, and a sense of regional belonging and shared identity.

The *Zwischenstadt* was IBA's solution to the same old problem of the Ruhr's lack of recognizable urban form and culture. As its directors returned to Schmidt's greenbelts as a way of shaping the landscape, they also returned to his intuitions about the possibilities of treating Ruhr's polycentrism as a strength to be promoted rather than a handicap to be overcome. Thomas Sievert's *Cities without Cities: An Interpretation of the Zwischenstadt* ([1997] 2003), published after Sieverts had served as scientific director for IBA Emscher Park from 1989 to 1994, elaborated the unusual urban vision driving IBA Emscher Park. The book, like the park itself, is another example of the Ruhr being "good to think with," and, like Habermas's *Structural Transformation of the Public Sphere*, it offered both a theoretical scholarly intervention and a guiding normative vision for urbanism, in this case focused on urban form rather than public life. In addition to providing a more favorable perspective on what planners had long understood pejoratively—and undifferentiatedly—as sprawl, the book also aspired to be a kind of manifesto for a new urban planning of the future, one that escaped the imaginative confines of the classical European city and laid out a set of principles for urbanism not governed by ideals of density and centrality. It was dedicated to Karl Ganser and heavily influenced by Sieverts's work in the Ruhr.

## Zwischenstadt

The word *Zwischenstadt*, which appears in German even in the 2003 English translation, describes what Sieverts calls the diffusion of "urbanized landscape or the landscaped city" produced by economic globalization, cultural pluralization, and the dissolution of city/nature boundaries. Instead of regarding "the urban periphery, the ubanized countryside, or the *Zwischenstadt* . . . as the essence of a lack of culture," the book "tries to increase awareness of this strange urban-rural landscape as a new form of city." Sieverts argues that *Zwischenstädte* (plural) can be recognized by features such as regional parks, corridors of built-up areas served by public transit, and the preservation of "non-hierarchical" nodes of centrality. He means for the concept to shed light on a real and growing number of urban regions that are "obscured by the myth of the Old City," arguing that *Zwischenstädte* are to be found all over the world, especially in places "where the historical, traditional, city-composing forces never took effect, such as in the Ruhr area, or also in the metropolitan areas of the Third World" (Sieverts [1997] 2003, xi, 88, ix, 12, 3). Foreshadowing the favorite statistic of the early years of the twenty-first century—that 50 percent of the world's population lives in cities (UN-Habitat 2007)—he predicted that, in the future, 50 percent of the world's population would live not in cities but in *Zwischenstädte* (Sieverts [1997] 2003, 11).

Sieverts's intellectual project was to understand emerging geographies of twenty-first-century urbanization using the Ruhr as a case: to offer a positive definition of regional landscapes without clear country/city boundaries and without the classical European city as the starting point and normative ideal—themes that are also of interest in contemporary urban studies (Brenner and Schmid 2015; Robinson 2006; Roy 2009) and that are close to the heart of this book. For IBA Emscher Park, as for Robert Schmidt in an earlier era, the concept offered a way forward, an urgently needed model for urban development in the region. As Sieverts quotes Karl Ganser in *Cities without Cities*: "The settlement belt of 800 square kilometers in the center of the Ruhr district is a built up area. With the weak growth potential of future times this *Zwischenstadt*—which neither complies with our image of the city nor with our yearning for an intact landscape—can no longer be remodeled. We must accept it as a given and endorse its hidden qualities. We must create order and draft images which will make this encoded landscape legible. This could lead to the development of a new type of regional plan" (Sieverts [1997] 2003, 121).

In economic and administrative terms, the concept was also meant to help prevent the city-region from disintegrating "into a series of selfish and competitive urban fragments" and promote "a feeling of belonging together" (Sieverts [1997] 2003, 60–61). In part, there was a sense that the Ruhr's best shot, economically, was to reinvent itself as a unified region to the benefit of all, rather than having individual cities compete with each other and work at cross-purposes. It also reflected trends toward the regionalization of spatial planning and economic policy found in Germany and throughout Europe in the late 1980s and the 1990s (Gualini 2004). For Christoph Zöpel, the former NRW minister who initiated IBA Emscher Park, the idea of "Metropole Ruhr" stands for "the importance of integrated city planning" and the "political integration of 53 cities" that the RVR oversees (Zöpel 2010, 48). For one of the city's mayors at the time (whom I interviewed, and who was also a planner who had worked on the IBA project), the decentralized structure is "one of [the Ruhr's] strengths, and one which is normally underestimated." Like Schmidt and the *Revierparks* planners in earlier eras, IBA once again revived the idea of the Ruhr as a region. "Essen for the Ruhr" was the geographic entity that received the European Capital of Culture award in 2010, "Metropole Ruhr" became the web address and headline concept for the RVR (RVR 2014b), and the RVR's public relations page described the importance of "regional marketing, with the brand name 'Ruhr Metropolis' as the core," which helps "put the polycentric conurbation in focus" (RVR 2014a).\*

For urbanists, what is most interesting about the concept is that it flips the relationship of city and countryside. When Sieverts described the *Zwischenstadt* as "urbanized landscape or landscaped city," by *landscape* he meant green areas traditionally understood to be not the city, typically called the country, or nature: agricultural regions, forests, parks, and other apparently natural areas. Much as Schwarze-Rodrian wanted to "turn it around," using greenbelts as the connecting spine of a regional urban landscape, Sieverts argued: "[In the *Zwischenstadt*] the ratio of open landscape and built-up areas has frequently been reversed; the landscape has changed from being an all-inclusive 'background' to being a contained 'figure.' . . . This *Zwischenstadt* is a field of living which, depending on one's interest and perspective, can be interpreted either as city or as country." IBA turned landscape (green space) into the glue of the *Zwischenstadt*, within which the settlements are intelligible because they "remain embedded as an 'archipelago' in the 'sea' of an interconnected landscape." It also turned the green spaces between municipalities from

---

\* While, by 2020, the RVR had a new website, the concept "Metropole Ruhr" remained (see Hospers and Wetterau 2018), and the new campaign for the Ruhr as a "City of Cities" retained the regional messaging and focus.

dead zones or background into "figure," making them part of urban space and, therefore, public, shared space (Sieverts [1997] 2003, 3, 9). In Karl Ganser's prescient words: "For 100 years, development [has] followed the logic 'town devours landscape.' For the coming 100 years, this will be turned around into 'landscape devours town'" (Ganser 2010, 9).

## *Industrienatur*

Turning the Ruhr into a park also involved a new understanding of industry, not as being synonymous with the destruction of nature, but as having created a special kind of *Industrienatur*. As an industrial region, the Ruhr was and still is commonly understood to be the antithesis of nature, having had its nature destroyed by industry. I have found no record of the origin of the word *Industrienatur*, which IBA used to describe the ecosystems that flourish in former industrial sites. But it used it alongside an older, sister term, *Industriekultur* (industrial culture or, more commonly, industrial heritage), that came into common usage as interest in industrial workers' culture and everyday life grew in the 1960s and 1970s (Berger, Golombek, and Wicke 2018). Rather than trying to fill the holes or erase the Ruhr's industrial history, both concepts celebrate its unique culture and ecology as the source of its contemporary interest.

Just as the *Zwischenstadt* rewrote the relationship between country and city, *Industrienatur* posited a new relationship between nature and industry. Consistent with their principle of "recycling," IBA's directors did not understand the Emscher Park project or the Emscher River renaturalization as efforts to return to an unspoiled first nature after over a century of industrial production. Instead, they conceived the Ruhr's altered ecology as a second, industrial nature to be deliberately showcased at ecological and cultural heritage sites. The Ruhr Museum, for instance—itself a former coal-washing plant recycled as industrial culture—designated anthropogenic sites as places of ecological value and turned them into places for recreation and objects of scientific study. It offered an outdoor "succession garden" with ruderal plants "under observation" (Godau and Heinrich 2010) as well as a large indoor exhibit on *Industrienatur* in which pressed, dried leaves and flowers were displayed suspended on a succession of walls constructed of glass cases in the style of curiosity cabinets. At sites where "wilderness" was allowed "to unfurl" (Weilacher 2010, 88), the flora and fauna that thrived were products of the environmental conditions that industry left behind. Many of the 244 hectares of preserved "industrial forests" (*Industriewälder*), for instance, are greened spoil heaps (the piled, discarded soil left over from mining activity) with gravelly, nutrient-poor soil hospitable to plants such as wild cherry, English oak, silver birch, and goat willow (Gausman, Weiss, Keil, and Loos 2007; Weiss et al. 2005).

IBA-produced parks such as the much-celebrated Landschaftspark Duisburg-Nord, located in Duisburg, another of the Ruhr's cities, turned *Industrienatur* into a design principle. As an article on the site put it: "The notion of *Industrienatur* attempts to reconcile two realms normally opposed to each other . . . by conceiving of nature and industry not as categorically different but rather as periods in historical succession, each becoming the foundation for the other" (Hemmings and Kagel 2010, 252). The park accomplishes this fusion of nature and industry by centering the deteriorating industrial past. It consists of the ruins of a former steel mill on and through which visitors can now wander and play, even rappelling off its outer walls and scuba diving in its gas tanks. Former interior rooms (now open to the sky) have become art installations and formal gardens; some have been flooded to create water gardens for aquatic plants. The landscape architect who designed the site told the *New York Times Magazine* in 2004 that "this situation is highly artificial" and "has nothing to do with untouched nature." He planned it to "[debunk] the fantasy of taking refuge in pristine nature" through a park that was "unmistakably manmade"—that preserved and made use of the site's existing industrial infrastructure and that offered the "attraction of honesty" by "deferr[ing] to the history of a densely populated and deeply scarred terrain" (Lubow 2004, 52, 49).

Narratives produced around IBA sites are explicit about this changed relationship. "In the middle of Europe's largest industrial conurbation," states a book titled *Emscher 3.0*, about the river's renaturalization, "[IBA Emscher Park produced a] specific postindustrial *Naturverständnis* [understanding of nature], in which city and nature, technology and art, industry and landscape, are no longer seen as fundamental opposites, but are in many ways amalgamated" (Wuppertal Institut für Klima, Umwelt, Energie 2013, 61; see also Frank 2010). *Emscher 3.0* goes on to describe the Emscher project as "tying together" the sanitation infrastructure of late modernity with modern concepts of the city and water and sanitation systems, putting "nature and landscape at the center of regional change, in a context where for the last two hundred years they [nature and urban development] have been opposed" (Wuppertal Institut für Klima, Umwelt, Energie 2013, 61). In this way, the concept of *Industrienatur* continues the blue skies promise of the 1960s—in fact, the subtitle of the book is *From Gray to Blue; or, How the Blue Sky over the Ruhr Region Fell into the Emscher*—by treating pleasurable environmental experiences and signifiers (blue skies, green grass) as symbols of hope for the future. But, rather than erasing the past, IBA recycled industrial history as a source of rich cultural heritage and unique industrial ecologies in order to tell a new story about the Ruhr. As Siebel put it: "The Ruhr has not had an identity of its own. Now that the industrial society which created and exploited the region is disappearing, the Ruhr has, for the first time, acquired a history of its own and thus it can develop its own special urban identity through its history" (Siebel 1999, 126).

## THE SOCIAL ORGANIZATION OF NATURE EXPERIENCE

Beyond settling on a vision, IBA's project directors also had to communicate it, using a variety of tools and techniques to project it into the built environment and into the minds of the public. They had a definite—if, as compared to the projects profiled in parts 1 and 2, much more abstract—set of beneficiaries in mind. Krupp's Margarethenhöhe was designed for an actual cohort of employees, and the *Revierparks* had at least a geographically defined universe of potential users. But an arts professional who worked in the Ruhr once told me that inhabiting IBA's visions was like being surrounded by "imaginary friends" for whom the landscape, museums, and programs were designed but whom one never saw. Two clear types of users were imagined: former factory workers and their families (longtime residents of the Ruhr) and outsiders (tourists from elsewhere in Germany or Europe). For outsiders, the goal was to create a new, favorable image of a place that had long been considered one of Europe's most polluted, proletarian regions as a place to visit, to invest in, and to which to bring your business. For locals, the goal was to "provide good quality of life" (Sieverts [1997] 2003, 74) and, as for the *Revierparks* planners in the 1970s, to make people "feel better [about deindustrialization], even though objectively the economic situation remains unchanged" (Ganser quoted in Barndt 2010, 277).

The Ruhr's unusual urbanism made getting people on board with this new story of hope and renewal especially challenging. Sieverts has compared the work of developing knowledge of the *Zwischenstadt* to coming to know a piece of music or a written text: "depending on their 'mode of reading,'" users "can combine fragments into different sequences and interpret them quite differently depending on their mood and experience" (Sieverts [1997] 2003, 109). This variability is true of any city, but, in regions like the Ruhr, which lack the morphology (symmetry, centrality) and iconic landmarks (the Eiffel Tower, the Brandenburg Gate) that lend themselves to simple, straightforward representation, interpretations become even more personalized, quixotic, and variable. As the urban planner Kevin Lynch put it in *The Image of the City*, a city's "legibility" or "imageability"—the "generalized mental picture of the exterior physical world that is held by an individual"—varies greatly by city type and has implications for how that city is inhabited (Lynch 1960, 4). In "highly imageable environments" such as, in Italy, Venice, or, in the United States, "parts of Manhattan, San Francisco, Boston, or perhaps the lake front of Chicago," one "would be well oriented and . . . could move easily" (Lynch 1960, 10), while "undifferentiated" Jersey City—like the Ruhr, a place where "the historical, traditional, city-composing forces never took effect" (Sieverts [1997] 2003, 3)—tests the relationship between the mental image and the actual space of the city (Lynch 1960, 10, 144–45).

My efforts to understand the Ruhr were consonant with Lynch's description of efforts to navigate undifferentiated urban environments. When locals guided me from one industrial monument to the next, they would decode what looked to me like perfectly ordinary streets, buildings, or open spaces, explaining that these brick cottages had been company housing, that this new conference center was built on a former factory site, that this apparent wilderness was actually a spoil heap only recently crowned with IBA art. But, when I set out alone on foot or by bicycle, I would find myself following a mysterious network of not-quite-finished roads that ended in hurricane fencing, trailed off into a field, or dumped me along a tangle of highway—and suspicious of my perceptions of the Ruhr's landscape. Was this a preserve for preindustrial wild nature or *Industrienatur*? Was this a portion of the renaturalized Emscher or just an overgrown sewage canal? Reading the landscape in these ways was made harder by industrial wastelands' tendencies to erase history and look timeless and ahistorical. Once production stops and infrastructure has been removed, cleared factory lots become meadows, coal heaps transform into forested hills, and rail lines turn into overgrown trails, leaving only the impression of an eerily pleasant and peaceful pastoral landscape. On some of these adventures, the Ruhr was a vision—Osthaus's glimmering, verdant city of the future. On others, it was an endless mess of housing, highways, and brownfields.

This was IBA's problem. The project directors knew that, to see the region as a giant park, people had to be able to experience it as one. And so IBA invested with the goal of "develop[ing] visions which open up the *Zwischenstadt* as a sphere of experience" (Sieverts [1997] 2003, 52), ensuring that people who traveled to visit the Ruhr had the correct experiences and interpreted them the way IBA intended. IBA, of course, produced new images of the region as part of its efforts—like Krupp's cinematic images, the *Revierparks* promotional brochure, and the Eisenheim movement's books and photographs—but, as Schwarze-Rodrian had insisted, they were not "going into cinema." They focused on physical interventions that, like Margarethenhöhe's decorative gardens, enabled and enforced new modes of interaction with the landscape. To render the Ruhr's complex landscape legible and replace its reputation of coal, steel, and beer with a new regional image of unique industrial culture and nature, the actual work consisted of the following.

## Experiencing Nature as Leisure

As in earlier periods, for locals to see the Ruhr as a giant park, they had to be able to experience former industrial areas as a leisure landscape. Residents were familiar with the Ruhr's parks and green spaces (it was only outsiders, who knew the region by reputation, who thought of the Ruhr as a place without nature), but its industrial spaces had—like agricultural land in the nineteenth century—long been spaces of labor. As in the 1910s and the 1970s, part of the greening of industrial spaces—in this case for recreation, green tourism, or cultural consumption—involved this transformation. Just as, in the 1970s, company housing was reimagined as a space of nature-rich alternative politics rather than oppression and control, and deindustrialization promised to make the skies over the Ruhr blue, reimagining the Ruhr's industrial wastelands still involved "dismantl[ing] industrial dominance, physically and psychologically" (McMullen 2007, 156). IBA needed to create new associations with an old landscape in a new historical period. An article on the Landschaftspark Duisburg-Nord described the shift as follows: "On a site where workers' movements were once closely controlled, and free roaming equaled transgression, buildings and structures now denote the exact opposite of labor and exploitation, namely leisure and recreation, as if it were a perpetual Sunday" (Hemmings and Kagel 2010, 251).

To accomplish this, IBA built infrastructure for new experiences of leisure. As a result of midcentury planning efforts—instituting a separation of functions and hiding industrial areas behind green curtains in the 1960s—and because many industrial sites had been secured and closed off when not in use, the land that became IBA Emscher Park was, though close by, often hard to access. To encourage people to venture into these *verboten* areas quite near their homes, and because IBA did not have the resources to acquire and remake every inch of the Ruhr, IBA invested in "injections" or "acupuncture": strategically placed museums, monuments, art installations, and points of scenic and historical interest (Hemmings and Kagel 2010, 247). These included projects such as the city of Bottrop's Tetrahedron, a steel pyramid topping an already-elevated coal heap that provided a new perspective on the region from an observation deck suspended fifty meters into the air, which always prompted remarks about how green the region looked when viewed from above. IBA also made these new scenic destinations accessible via hundreds of kilometers of bike lanes, roads, walking paths, sports fields, waterways, and meadows (along with accompanying maps, guidebooks, visitor centers, and guided tours), investments that Schwarze-Rodrian described as a "service to the daily user." They deliberately located access points near the new social housing in which IBA had also invested, and they designed the routes to enhance the sense of the Ruhr as a region by creating new connections between adjacent cities and sites.

The goal of these landscape interventions was to develop residents' "consciousness of being an inhabitant of a holistic city region" and "create a new relationship to 'place' that strengthens identity and provides support" (Sieverts [1997] 2003, 60–61). When I asked, somewhat cynically, "What do you think these park spaces actually *do*?" Schwarze-Rodrian answered without a trace of sarcasm: "They qualify the conditions of life. And that's a very wonderful thing. They are very easy to understand. . . . You know about our bridges? We have concentrated very strongly on saying, 'Let's build bridges in the Emscher Landscape Park,' because bridges are so important for connecting urban landscapes. . . . The benefit of urban landscape is that people love it, they use it, they use it for sport and immediately say, 'That's my neighborhood.' It doesn't even take a day before they say, 'And my neighborhood is so and so.'"

IBA commissioned artists and designers to repurpose railroad bridges or build new pedestrian and bicycle bridges over roads and former sewage canals to connect residents to the new recreational areas and link up portions of the park. After explaining the design of one bridge in loving detail, Schwarze-Rodrian concluded by saying that well-designed bridges provided "new interactions" with the landscape: "The Emscher Landscape Park is decentralized. That means it's in front of your house. It is not only [Landschaftspark] Duisburg-Nord and Phoenix Lake. This road system has several hundred kilometers now, and they did it all in the last ten, fifteen, twenty years. And in that way it is not exclusive. It's accessible. Accessibility is a very important idea." The goal of these physical investments was to help users understand the region as a leisure landscape and the park as a regional resource that belonged to them by leading them beyond the segment in front of their house.

## Soft Management

Another task was to encourage a sense of ownership and buy-in among the city administrations, which IBA accomplished by helping representatives of each of the Ruhr's cities experience themselves as part of a whole. For this IBA relied on a principle Schwarze-Rodrian described as "soft management." Just as Emscher Park was "not something to own," with no land ever formally secured, no central management body for IBA Emscher Park was ever created. Schwarze-Rodrian explained this as follows:

> The whole park is an informal thing. There's no regional park administration, no Department of Parks and Recreation. . . . It would make things easier in some ways [if there were], but in other ways it would make things stupid. The Emscher Landscape Park is used by the people, day by day, and the RVR, in 2006, after the Master Plan, got an additional paragraph in its law, to develop and maintain it as regional park system. They took it seriously and declared the RVR to be the ongoing management body. So there is nothing central. It's decentralized. It was from the beginning and stays so today. So Mr. Sierau [the mayor of Dortmund] is doing things in Dortmund, it happens in the Emscher Landscape Park, and we discuss and interact regionally to see if it is fulfilling [the IBA vision,] ideas, etc. . . . This is soft management, not a hard top-down thing.

Toward these ends, the IBA directors instead saw their role as facilitating coordination between cities. Schwarze-Rodrian described his job and the task of the overall Emscher Park plan as "mainly moderation." "I was not the director of the park," he said. "I was the moderator of the park." As a moderator, he saw his responsibility as "just opening, making it possible for people to start to rethink reality and develop things together." His basic procedure was to identify sympathetic allies in each of the seventeen cities, people "engaged in urban landscape quality," and then "bring the neighbors together" in order "to create an operative understanding of urban landscape development." He created seven working groups around each of Schmidt's seven green corridors, each of which would "work together in their neighborhood": "The landscape [of the park] lies just between them, at the edges of the single cities. And it is just in these areas where the river is. The nature, and the water, is very banal—it is not where the city center is, etc. So to develop these things we have to find new responsibility between neighbors. Experts in each area only know about their own area but not other stuff, even sixty kilometers away."

The actual labor of getting people to work together was also quite banal. Schwarze-Rodrian engaged in a variety of typical facilitation tasks for IBA's city partners over the ten years. He organized meetings, commissioned projects, wrote reports, synthesized findings, and held what he called the "golden reins"—distributing €500,000 in state money to local working groups over three years, most of it going to planners whose job it was to collect the groups' ideas and turn them into concrete plans. Such incentives for participation got professionals from each city working together on specific projects and set a precedent for cooperation within the region. IBA's "acupuncture" approach also encouraged participation. Locating projects on the borders between cities made people in different municipalities work together (just as it encouraged residents of different cities to visit the sites) by having pairs of cities develop projects that fit the central guidelines (Hospers 2004, 153).

Longer term, IBA envisioned a shift from "Microsoft to Linux": from centralized, top-down project management to decentralized (though still basically top-down at a smaller scale) planning. Karl Ganser embodied the Microsoft approach. According to Schwarze-Rodrian, Ganser had been a "total manager," one who was "able to define the conditions of the whole." But after IBA, he explained, "there was no state [acting as a] creative moderating figure ... no hidden director in the Ruhr." Instead, loose collaborative frameworks allowed individual cities to carry out development projects in keeping with the IBA vision and in cooperation with their neighbors. The principal tool for this collaboration was the Emscher Park Master Plan developed in 2010 by Schwarze-Rodrian under the auspices of the state-funded Projekt Ruhr GmbH after IBA ended (see Projekt Ruhr GmbH 2010). Schwarze-Rodrian says that the master plan "is a contract between twenty cities, and not a contract with a government," in which, as in Linux, "the users make the rules, instead of the software owner."

## Promoting Industrial Culture

To reduce variation in experience, and to encourage people to interpret variable experiences as intended, IBA also used the strategically located "injections," serving as both symbolic icons and geographic anchors, to create standard ways of perceiving and moving through the region. Achim Prossek, a geographer who has written extensively on image marketing in the Ruhr, has pointed out that IBA explicitly compared its new industrial monuments to the great landmarks of more "legible" European capitals (Prossek 2009). Zeche Zollverein, for instance, closed as Essen's coal-washing plant in 1981. In 2001, it became Germany's third UNESCO world heritage site and renovations began to turn it into the Ruhr Museum with a master plan designed by Rem Koolhaas. The renovated building became known as "the Cathedral of the Ruhr region" and its pithead tower a symbol of structural change (Prossek

2006). When IBA renovated the cavernous, 384-foot-high Gasometer in Oberhausen as an art and exhibition space in 1994, a poster printed for its opening announced, "Welcome to the club," and pictured the Gasometer next to the Eiffel Tower, the Cologne Dome, and the Brandenburg Gate. This was, Prossek says, "the first time an industrial relic was seen as an icon for a city" and a "cultural and historical equivalent to other well known iconic buildings of European cities" (Prossek 2004, 68). Monumentalizing such sites was part of IBA's elevation of working-class culture to high culture, but it was also a way of increasing legibility through simplified, recognizable images of the region. These sites also helped standardize routes through the Ruhr. Without a large central city, visitors had no idea where to begin their journeys, and IBA hoped that the Ruhr Museum at Zeche Zollverein would serve as a "gateway" to the region (Prossek 2006).

IBA also used high-profile art, architecture, and cultural events to promote the new regional image and pride in and curiosity about the region. These efforts were mainly directed toward outsiders and focused on putting *Industriekultur* on display. In addition to making great large-scale exhibition spaces, industrial buildings are also visually and acoustically unusual and, thus, make dramatic settings for theater, music and light shows, festivals, and other events. During IBA Emscher Park and after, IBA-organized festivals and events promoted industrial relics as music and theater venues, boosting their appeal by creating a sense that they were spaces touched by magic and mystery. A Czech perspective on the IBA (Slach, Rumpel, and Boruta 2011, 214) described this use of festivals and staging as creating spaces and times of "out-of-everydayness" (*Ausseralltäglichkeit*) that inspired feelings of inhabiting sacred time and collective effervescence—Durkheim's term for the energy and moral unity generated through religious experience (broadly defined).

## Denaturalizing Industrial Nature

For industrial nature, the challenge was to make sure that natural-*looking* landscapes were seen as the products of high-tech engineering they were, while also making industrial sites recognizable as "nature." For example, referring to the ecological restoration of the Emscher River as a *renaturalization* is in almost all senses a misnomer. As IBA and the Emschergenossenschaft (the Ruhr's regional water management body since 1899) restored the river system's ecological health, they had to re-create the river's aesthetics of naturalness because the original bed and tributaries had been so completely "rationalized" and "canalized" during over a century's use as an industrial sewage canal that there was no recognizable first nature for the river to return to. Though there was some discussion of preserving sections of the rationalized and canalized river (a concrete-lined V) as cultural artifacts more akin to *Industriekultur*, the Emscher was reconstructed with soft edges, wild grasses, and a gently curving path—all completely artificial—while signage, promotional materials, and commissioned public art projects showcase the four hundred kilometers of underground wastewater pipes that make the natural-looking river possible (Huning and Frank 2011). Experientially, the new Emscher's aesthetic of naturalness is visually consistent with other urbanized nature projects examined in this book, in that it uses green space and vegetation to create an environment meant to be experienced as pleasant, pastoral, and idyllic.

For people who live along sections of the river that have not yet been renaturalized, such changes are unimaginable, while, for those who live near already-restored sections, the Emscher's history is easily forgotten. To ameliorate both conditions, the Emschergenossenschaft also produced materials depicting the river in the past, the present, and the future. In 2012, for instance, it distributed small plastic cameras as promotional giveaways. When you looked through the viewfinder and pressed the shutter button, you could flip through a series of before-and-after images of the Emscher, first in its rationalized state (straight lines, concrete walls, fenced off or inaccessible, edged by manicured lawn), and then renaturalized (curving gently, with tall reeds and grasses growing along its banks, trees planted in the background, and sometimes a wading bird or a relaxing human in the foreground). In the same year, the Emschergenossenschaft had installed a larger-than-life version of this experience on the lawn of Dortmund's Zeche Zollern colliery and industrial museum. Visitors were invited to enter three shipping containers labeled "yesterday," "today," and "tomorrow." Inside, each depicted a phase of the Emscher's history: industrialization, rationalization, renaturalization. The LCD screens lining the inside of the "tomorrow" container projected images of birds, animals, water, and sunsets and the sounds of birds singing and water running.

## THE ERASURE OF THE SOCIAL ORGANIZATION OF EXPERIENCE

Experiencing IBA Emscher Park's landscapes as nature did not require complete escape from society or an apparently untouched wilderness. What is intrinsic to the popular understanding of a nature experience—and therefore to urbanized nature—is the perception of an *unmediated* experience. To take an example from another context, a number of years ago a pair of red-tailed hawks were found nesting in New York City's Washington Square Park. Though they were living in a human environment and visible only as a result of human intervention, fans who watched them on a live-streaming hawk cam placed near the nest became upset when signs of the social—a hand, a plastic bag—intruded and broke the frame of their view (Angelo and Jerolmack 2012). In a similar way, in the Ruhr, all the social work surrounding greening—coaching people to experience the landscape in a particular way—coexists with a narrative of creating opportunities for people to experience that landscape as unmediated. IBA planners described their efforts and technologies as just "facilitating" or "opening up" the Ruhr's existing landscape for people. We will see in the next chapter how audiences receive these messages, but it is notable that the greening protagonists who are engaged in this imaginative work and who are actually constructing these projects describe their actions in these terms.

Haiko Hebig, a landscape photographer and lifelong Ruhr resident I interviewed about IBA Emscher Park, describes IBA's strategies, critically, as "viewing instructions." He sees them as misleading projections of an imagined world that cannot be brought into being, one focusing on an aspirational future at the expense of the history and struggles of the past. He makes his own photographs that critique IBA's projections by establishing a different relationship to the deindustrialized landscape; rather than creating aspirational images that bring a future into focus, they denaturalize and decode apparently natural landscapes to make the past visible. In a 2010 photograph called *Hochofen Phoenix 3 (No. 3 Phoenix blast furnace)*, for instance, the Phoenix factory site described at the beginning of this chapter has been cleared of buildings, Chinese laborers, and subsequent rubble and has become a level field set up for a carnival (see fig. 5.2).

In the middle ground of the photograph, the only human in the scene is peering at a bulletin board. A detail of this portion of the photograph that Hebig provides reveals that the man is looking at IBA's images of industrial buildings as the lit, monumentalized sites they are to become. Hebig pairs this image with this caption: "To make sure you know how to look at the scene, some brave new world photos and visualizations were shown on-site." In other work, he provides a corrective with alternative views, retrospective instead of prospective. A photograph of the placid, artificial Phoenix Lake eventually constructed on the site is titled *BOS Steelmaking Shop*, while an image of late-afternoon sun stretching across a meadow is called *Wulfen Coal Mine*. Unlike IBA's projections, Hebig's photographs are meant to remind viewers of an industrial history still too easily made invisible in spite of its ongoing felt effects,

FIGURE 5.2. Haiko Hebig, *Hochofen Phoenix 3 (No. 3 Phoenix blast furnace)*, 2010. (Courtesy of the artist; https://hebig.org/photos/beispiele/ho3/.)

but, like IBA, Hebig, too, is mediating experience, in this case framing sites of apparent nature as having industrial history.

Hebig is also right, of course, to remind us that projecting a future in images or words is not enough to make it a reality and that changing the perception of a region is not the same as retraining its workforce or providing affordable housing or building new industries. And IBA received such criticism. Some thought, for instance, that the very goal of changing an image was a frivolous use of resources in a region with such pressing economic needs. And, as we have seen, although IBA was conceived as part of the project of shifting the Ruhr to a new economy, Ganser and others acknowledged that its economic impact was not as significant as its effects on peoples' feelings.

I take up Hebig's phrase here not so much to criticize IBA's message or goals as to point out how ubiquitous this kind of projective work is across projects and places. Greening projects are technologies of experience and impression management. They lay out ways to interact with the environment that cause people to experience places and people in particular ways, and they offer narratives of that environment—including, sometimes, the idea of society as separate from nature and, therefore, of green space as an escape from the social world—in the forms of engagement they lay out. IBA's activities were unusual only in that it is rare to have so many resources dedicated to imaginative projects and to have protagonists so self-conscious and reflexive about their activities (and, being academics, reflect on them in writing). But they bring into relief practices that are characteristic of greening more generally. Sieverts writes: "The work on the sensitization of inhabitants to the *Zwischenstadt* and with it the work of positively influencing mental images is a task which is as diverse as it is charming and unending" (Sieverts [1997] 2003, 107). He conceives of the city "as the common product of the 'hardware' of the real, physical environment and the 'software' of perception and use," drawing on Kevin Lynch to argue that "the interactions between both worlds is [*sic*] what allows 'city' to exist in the first place" and that "*the most effective way*" to alter the city is "*to address the interaction itself, by directly involving the inhabitants in the reorganisation*" (Sieverts [1997] 2003, 101). Put in more sociological terms, creating a new imaginary of the region was as much a product of the tools and technologies used to provide a new frame for experience as it was about physical transformation. These activities were especially important in a *Zwischenstadt* because of its illegibility, but they are necessarily part of any citymaking project, not only in *Zwischenstädte*, and not only green ones.

The people involved in this work—IBA's project directors in this case—were well aware of all this. However, after Schwarze-Rodrian's many efforts to connect people to the park, the indicator of success he looked for in the end was their disappearance in the minds of the public. He concluded:

> But for a normal visitor all this is not interesting. You want to do things, you want to lie in the sun, you want to bike, you want to [experience] good [environmental] qualities, etc., you don't want to be involved in discussions like this. This is only for professionals. . . . [We make] great, great infrastructure investments, in environmental quality, and [especially in] parts [of the project] that you cannot see—to exchange [dirty water for clean, i.e., to maintain sewage functions] in tubes underneath the ground, that are not seen! I don't know if you've ever renovated an old house. To make a wall flat—spackling—can be a hard job. If you successfully make it flat and put wallpaper over it, then nobody talks about your wall. If you have buckles, etc., then they [will comment on it]. So you work on it to be perfect, and the result is that nobody sees it. You know it. I'm talking about renovation. So the transformation includes renovation. Renovation is not so visible, not so attractive like a new building, but you shouldn't be sad about that. You should know that.

Schwarze-Rodrian described IBA's work and actions not as a normative project, or as imposing a civic vision on people, but as providing access—in his words, as "making possible," as "facilitating," as simply offering an experience of something already there. To put this differently, for all that I have characterized greening as an aspirational social practice of creating urban futures, those involved in the process do not ultimately describe their own work in these terms. IBA's directors saw themselves as creating infrastructure and opportunities so that they could then step back and allow people to have their own nature experience qua unmediated experience, even though all involved were fully conscious of these environments as a human-produced industrial nature rather than first nature.

This dynamic—the erasure of the social organization of experience—is an effect of urbanized nature. The words of greening protagonists suggest that making a nature experience synonymous with unmediated experience is even more important than the specific phenomenological forms that greening projects take. Such mundane acts as creating stories, images and marketing materials, and consistent routes and experiences are necessary to create experiences that help frame places in particular ways and stabilize those frames. But, if successful, all that staging work is not part of audiences' perceptions of the projects: they should not be conscious of themselves as being managed in these ways. This pattern also contributes to the characteristically paternalistic nature of greening projects because, as the next chapter shows, in spite of greening protagonists' awareness of their own efforts, the erasure of this framing work allows them to understand themselves as acting benevolently and in the public interest—as simply providing public resources and access to public goods— rather than exercising forms of control by projecting futures and organizing experience in particular ways. It also makes it possible for audiences to receive these spaces as public goods rather than as acts of managerialism, in spite of the fact that users are always receiving and responding to guidance about how to view and interpret green spaces, including and perhaps even especially being directed to view them as nature.

CHAPTER 6

# Experiencing Nature as a Public Good

I moved to Dortmund in 2012, more than ten years after IBA Emscher Park's official conclusion. Dortmund is one of the Ruhr's larger, medieval Hellweg cities, with about 500,000 residents, a traditional urban core, and a local economy stronger than those of the younger industrial cities of the northern Ruhr in which IBA had focused its efforts. But it was also the location of the Phoenix plant, the one that had been sold to China and had left the giant hole in the neighborhood of Hörde, which had once been the primary residence for many of the steel- and ironworkers employed at the site.

When I arrived, the site and the city were in the midst of quite a significant transformation. In 2009, one of the planners involved in the IBA project, Ulrich Sierau, became Dortmund's mayor. One of his goals was to extend the "green ring" of Emscher Park around Dortmund—linking it up to the regional landscape park—which dovetailed nicely with a local architect's ten-years-in-the-making dream of turning the Phoenix site into a giant artificial lake. In 2010, the site had been flooded to create a lake three-quarters of a mile long and just over a quarter of a mile wide, which was surrounded by public parkland and two surfaced paths for walking, biking, and rollerblading. Beyond the public space, thirty-one hectares of new, single-family houses (Wuppertal Institut für Klima, Umwelt, Energie 2013, 83) made up what was at the time the largest block of middle-class housing in Dortmund. Phoenix Lake was just half of a former industrial complex that occupied two sites lying, respectively, east and west of Hörde. Phoenix East, where the lake and the housing is, was once the location of a steelmaking, -casting, and -rolling factory. At Phoenix West, at the time being redeveloped for new commercial uses, iron ore was once prepared and coke and pig iron made. In 2012, both parts of the project were still under construction, but they were being marketed under the slogan "Wohnen am Wasser, Arbeiten im Park" (Live on the water, work in a park). The city hoped the redevelopment would help rehabilitate Hörde's neighborhood economy and attract new technology and manufacturing companies—and their employees—to Dortmund. Employees could "live on the water" by purchasing land and building a large, contemporary house overlooking the lake at Phoenix East and "work in a park" by biking along the wooded trail to Phoenix West, the site of new office buildings and converted industrial spaces.

In 2012, the lake was a wildly popular new public amenity in the city, and at first glance the Phoenix project looked like the success of IBA's vision: the extension of *Industrienatur* as an economic development model, the rehabilitation of industrial wastelands as recreational spaces, and the take-up of a green, regional vision for the Ruhr. The redevelopment itself—a lake, green space, recreational opportunities—was materially in keeping with IBA projects. Phoenix West's remaining industrial infrastructure was just the stuff of IBA's paradigmatic images of *Industriekultur* and was, in fact, *exactly* the same stuff of two of IBA's most iconic landmarks: a blast furnace like that at the Landscape Park Duisburg-Nord and a Gasometer, a gas storage tank whose twin in Oberhausen had been converted into the cathedral-like, one-hundred-meter-tall art gallery. Michael Schwarze-Rodrian described Phoenix Lake as part of the post-IBA trajectory from Microsoft to Linux: from centralized, top-down project management to decentralized (though still basically top-down, though smaller-scale) planning.

But I was surprised to learn that public sentiment was divided and that at least some local officials had a very different perspective on the project. Largely as a consequence of IBA Emscher Park, today the Ruhr is publicly upheld as a region that has embraced its industrial past. In comparison to cities such as Glasgow, Dortmund is seen as a place that acknowledges its industrial history, has preserved its industrial relics, and makes that narrative part of its contemporary economic development (Richter 2017). But this story is not an uncontentious one. Though the city's mayor—having been part of IBA—offered such a narrative of Dortmund and the Ruhr, I was told by one urban professional that one should "never say in Dortmund that Phoenix is an IBA project." Local professionals and private citizens offered at least one competing way of narrating the Phoenix project and of understanding the city. Rather than describing it as part of a unique urban region that had embraced its industrial past, local professionals were working hard to pitch Dortmund as a more conventional midsized city that saw itself in competition with cities like Hanover or possibly

Vienna rather than Paris or Berlin. The performance of nature at Phoenix Lake reflected these latter ideals.

Chapter 6 shifts perspective from providers of nature to their target audiences in order to examine the dynamics of greening projects' reception and especially how urbanized nature makes it possible for them to be received as public goods. It identifies a recursive quality in how greening projects are discussed and debated in two sets of receiving audiences, employees of the city of Dortmund (IBA's audience) and members of Dortmund's public (the city's audience). Both first critique nature—like greening protagonists, they are absolutely able to describe the projects' problems and shortcomings (i.e., to see them as social projects)—but then go on to advocate for perceived-to-be-universal nature in ways that reproduce the practice's usual blind spots and aspirations. In addition, the collective love of nature gave public critiques a specific form: one that was slow to emerge and that never fundamentally questioned the idea of nature as a public good at all. More concretely, those involved were able to maintain their faith in nature even in the context of imperfect projects by carefully carving off the untarnished "nature" (in this case, the lake) from the projects' necessary compromises and superficially "social" aspects (in this case, the housing surrounding the lake). And, though receiving audiences eventually critiqued failed efforts to isolate nature as a public good, following their critique they, too, went on to make arguments mobilizing nature anyway.

This is the social consensus of nature's universal benefit in action and an illustration of how the moral weight of everyday forms of nature affects planning, politics, and public discourse. On the part of greening protagonists, urbanized nature allows well-intentioned action: it makes it possible for people to act in the name of the public good through nature. In the world of reception, meanwhile, both greening protagonists and receiving audiences are able to sustain this belief even in the face of constraints: receiving audiences tend to accept greening projects as public goods even as they fail to serve the public equitably. Urbanized nature contributes to these dynamics by shaping the public discussion surrounding and the material outcomes of greening projects. This social imaginary forecloses discussion of these projects *as social projects* and, therefore, of their possible benefits or negative effects.

### REJECTING A PUBLIC GOOD: WHY IBA'S VISION FAILED FOR DORTMUND

The conflict surrounding IBA's legacy in Dortmund was a product of the shift from generously funded IBA projects to municipal-level business as usual, a situation in which the ideal of providing of public goods was inevitably grounded and compromised by practicalities. Motivated by the idea of nature's universal benefit, IBA saw itself as giving a gift to the Ruhr. But at least one receiving audience had to confront the social reality of IBA's projects, in spite of its view that it was simply opening up the landscape: Dortmund's municipal economic development professionals. As IBA's significant framing efforts revealed, it was not simply offering unmediated nature as a public good. On the receiving end, for a city like Dortmund that lacked both the ample funding and the powerful regional management body that made IBA possible, living with the projects proved to be far more complicated.

If you follow a shaded gravel path at the far end of Phoenix Lake along a small creek, past the old railway bridge that once carried iron from Phoenix West to the steel mill at Phoenix East, you will find yourself in Phoenix Park, as the sixty hectares of brownfield surrounding Phoenix West are called. The site, adjacent to a towering blast furnace, with some (at the time) mostly empty office buildings along its edge, is mostly designated as an official "nature reserve" (*Naturschutzgebiet*) that is part of the Emscher Landscape Park and home to, among other things, an unusual species of Alpine frog that the reserve is intended to protect. The park is actually quite beautiful, and large, but Phoenix West has been a more complicated and difficult kind of site than crowded Phoenix Lake just half a mile away. On a cold, windy day atop the blast furnace, while escorting me through a maze of metal gates and passages built for maintenance, for security, and occasionally for carrying visitors through the structure, Wolfgang Meier,* a Ruhr native and recently retired employee of the city of Dortmund's economic development agency, recounted redevelopment challenges that helped me understand why IBA's vision was simply not tenable as a local economic development model and why its message and guiding concepts were effective only within the context of a regional economic and administrative state of exception.

Many of Meier's remarks underscored just how much more financially constrained Dortmund and other Ruhr cities were than IBA and how financial realities complicated the projects, even in terms of just receiving and maintaining these so-called gifts. The blast furnace I was standing on top of—on a metal catwalk, trying not to look down—was Meier's pet project, and, as we were buffeted by the wind, he explained that *Industriekultur* and *Industrienatur* require massive investments that often come with little economic payoff.

---

*With the exception of Dortmund mayor Ulrich Sierau and the photographer Jürgen Evert, all individuals quoted in this chapter are referred to by pseudonyms.

The city had hoped to rehabilitate the blast furnace as some kind of commercial space or publicly accessible form of *Industriekultur*, but it was finding this far more difficult than had been anticipated. When I asked whether the Landschaftspark Duisburg-Nord had been a model for Phoenix West, Meier explained that a Duisburg-Nord was impossible, in spite of the soon-to-be-crumbling industrial infrastructure and existing park-like quality of the site, because Dortmund had learned from IBA exactly how expensive it would be to maintain it. The park had been built by IBA with public funding and then donated to the city, which at the time was "very happy about it." However, Meier continued: "If you ask the city of Duisburg now, they would say, 'We'd never do that again,' because it's awfully expensive to run it." As an example of the hidden costs of maintaining even apparent ruins, he pointed to a partially collapsed adjacent building that had been designed for heat-generating interior uses (making coke and pig iron), explaining that the roof had fallen in the year before after a heavy winter snow because there had been no heat inside to melt it away. Even if Dortmund had had the maintenance dollars, there was no funding to redevelop the site as an industrial heritage destination anyway. NRW Urban, a private subsidiary of the regional government that was helping redevelop Phoenix West, had tried to argue to the state of North Rhine–Westphalia (the same regional government that had funded many of the Emscher Park projects) that the blast furnace and gas tank should be preserved as tourist destinations, but they had failed. So, in 2012, the blast furnace—which Meier said was "too dangerous" to be left open and "too expensive" to be retrofitted for free public access—was closed to the public most of the time, except on tours occasionally organized for school groups, potential developers, or visiting sociologists.

In the absence of state funding, the city and NRW Urban decided to invest a minimum of funds to stabilize the blast furnace and remediate the land and try to sell it to private developers—for €1.00, in exchange for taking on responsibility for maintenance. Here again, practicalities stymied the projects. While IBA had to answer to no one before turning Essen's coal-washing plant into the Ruhr Museum, Dortmund had applied for EU funding to stabilize the buildings at Phoenix West and develop new adjacent office space. That funding stipulated that occupants must be new businesses attracted from outside the area rather than local ones. Because finding such businesses was proving more difficult than expected, the new office space stood partially empty. The city was also having trouble finding investors for the blast furnace because, in Meier's words, it was not a "normal real estate project." It "didn't fit" the usual formula of "good location, good environment, good accessibility," and, worse, came with an estimated upkeep of €150,000 per year. Meier had looked hard for potential buyers, but even he reluctantly admitted that it was "good for nothing," explaining: "It's not a building. It has a roof, but no walls. . . . It's always windy, and there's no sun." A group of restauranteurs had declared the site unfit for even a beer garden.

Moreover, Phoenix West could not be another Duisburg-Nord—even if the city had the money—because municipally organized economic development meant that there was competition between cities and, as a result, a problem of repetition. The very existence of Meier's job and the responsibilities with which he was tasked—to redevelop Dortmund's industrial wastelands for new economic uses—belied the fact that the city functioned as a competitive economic unit. As part of a regional, tourism-based economy, IBA's *Industrienatur* and *-kultur* were intended to attract people to the Ruhr for a long weekend or a holiday visit. But as a regional destination the Ruhr simply did not need two Duisburg-Nords, and, for municipal economic development, such landmarks added little to the calculus of where to locate companies. "There really can't be a second Landschaftspark Duisburg-Nord," Meier said, because the reality of intermunicipal competition required making Dortmund stand out from other Ruhr cities rather than emphasizing similarities between them.

Meier described representational as well as economic problems with the IBA model. I knew from my own experience that, in spite of IBA's framing efforts, the Ruhr still lent itself to widely divergent experiences and interpretations. Conducting interviews at Phoenix Lake, for instance, I had heard from one individual that the site was undesirable for young people because it was too far out of town, while the next had said that he would never live there, "in the center of the city." And, while such indeterminacy is in harmony with the character of the *Zwischenstadt* and was appealing to the intellectually minded IBA, whose objectives were not primarily economic, it was a disaster for city marketing.

On what I thought was quite an interesting walk around Phoenix Park after we had toured the blast furnace, Meier paused, sighed, and remarked: "In winter it's rather dull, this place here." He went on to explain how challenging it is to translate *Industrienatur* and *-kultur* into a viable local economic development message in a city that has "very, very few things" to boast about. In my interpretation, his comments suggested that these concepts are not iconic or stable enough signifiers for the region. As he put it: "You can't transport *Industriekultur*. You can show people *Industriekultur*, and they [will be] overwhelmed by it when they see it, but you must see it." For instance, Meier argued that it was still essential to introduce people to the city by guided tours:

You must get people here, you must hike around with them. When we get people here from the outside who are trying to decide on a city to relocate jobs, we talk to them about the different companies here, about business connections, about living in Dortmund, about Borussia [the local, beloved, and very successful national soccer team], and then we take them in the car, and drive to the Ruhr mountains and to Phoenix Lake, and hike around. Sometimes I take them to the Dortmunder U [a brewery redeveloped as an art gallery, movie theater, and bar]. And, well, then they say it's an interesting city. They've got me as a pathfinder, you know?

Following this remark, he asked if I would still be in Dortmund in the summer. When I answered affirmatively, he provided some instructions about how I might go about finding the beautiful parts of the city without him as a "pathfinder." His instructions were enlightening given my own previous failures. I needed to strike out at the right time of year (June), by the right means (a bike), and on the right route (starting at Palmweide and then biking all along the Emscher, to Phoenix Lake, on to Aplerbeck, and then to Holzwickede). "You will see," he promised. "It's very beautiful." Then he went on:

It's a special situation. This industrial area has many very beautiful parts, but the parts are very small. You must know them, and, if you don't know them, you're lost. I always remember [this story]. Friends of mine [once came to the Ruhr], and we hiked around with bicycles for the weekend. And then they went back to the south of Germany—they live in an area where people go for holidays, you know—and they were so interested in what they saw here that they said [to their friends], "Well, we've just been to the *Ruhrgebiet*, and boy it was interesting! We had a weekend, and it was overwhelming." And so three girls said, "Well, we'll do the same." And they took their bicycles and got a hotel in Essen and came here for a weekend, and then they were bored, absolutely bored. They arrived back in Würzburg and said, "Why didn't you tell us it was so boring?! The *Ruhrgebiet*—I've never been to such an ugly place as that!"

Michael Schwarze-Rodrian might have found this story sobering. I felt validated by it, having until this point chalked up my own intermittent inability to see the Ruhr to a lack of imagination. "And so they didn't know, they couldn't find it?" I asked (*it* being the elusive beauty and pathos of the Ruhr-lover's Ruhr). Meier answered: "They couldn't find it, yeah. You must know the places, and then it can be very interesting, but as a whole it's an industrial region, and it's ugly. That's the way it is. You must *know*." To me, the story suggested that, in spite of IBA's efforts, there remain far more possible ways to experience the Ruhr than its directors would have liked to imagine.

For municipal economic development professionals, the lesson of Meier's account is that you cannot afford *not* to look a gift horse in the mouth—and that, when you do, the gift does indeed turn into something else. As an individual, Meier loved *Industrienatur*. He was considering taking on Phoenix West's rehabilitation as a volunteer in his retirement, partly driven by the conviction that—as with the Ruhr itself—people would be really moved if only they could experience it. Our interview was punctuated by my "wows" and exclamations as we climbed around the giant structure, and Meier (agreeing with IBA here about the power of new perspectives) said that visitors seeing the aerial view of Dortmund for the first time "look like little children, with big eyes, going whoa!": "It's overwhelming." His account made clear that Dortmund's rejection of IBA was due not to a lack of vision or understanding on the part of rigid city technocrats but rather to the fact that *Industrienatur* simply did not translate into a feasible local economic development strategy. A park like Duisburg-Nord seems like a welcome public resource until you look at its maintenance budget. Repurposing a blast furnace like the one at Phoenix West seems like a great idea until you try to fund-raise for it or contend with competition from the one in Essen.

For us, the lesson is that the fantasy of urbanized nature breaks down in the realities of day-to-day experience. Even as these projects are carried out and gifted as public goods, for people involved with these sites, more complicated realities make it impossible to view them as either universal or unmediated, much as it is hard for a farmer to view an agricultural landscape as pastoral or idyllic. Though IBA saw itself as benevolently offering nature as a public good—and perhaps also a replicable model for doing so—it was not a lack of imagination but practical concerns regarding maintenance, competition, and marketing that made it impossible for midlevel employees (if not the mayor) to take IBA on as a model for local economic development. The city broke from IBA, not because it could not see the beauty of the Ruhr's industrial past or did not believe in *Industrienatur*, but because of practical constraints and economic imperatives. As much as Meier wanted to save the blast furnace, he had to find tenants, raise money, and sell the city on the idea. All these were effects of the shift from a regional project operating in a state of exception to one abiding by the local demands of municipal business as usual.

## THE NEW DORTMUND, NATURE AS NORMALIZING

However, Dortmund's experience with the realties of IBA sites, its vision of the Ruhr as a unique region, and its critique of *Industrienatur* as a local economic development model did not make the city administration any less able to draw on greening as a mode of urban improvement. In fact, just the opposite was the case. In spite of its rejection of IBA, the city took up the same discourse of nature as a public good at Phoenix Lake and used the site to communicate its own urban vision, just as have all the other protagonists in this book. Ulrike Czerny, the press officer for the Phoenix Lake Development Corporation—who, like the mayor, had also once worked for IBA Emscher Park—described Phoenix Lake as a definitive break from IBA. While she agreed that IBA had helped "unify the cities," produced "new partnerships," and succeeded in "transporting" the idea of the Ruhr as "one big region," she also told me that the feeling in Dortmund was that the region was oversaturated with *Industrienatur* and *Industriekultur*, remarking: "I think everybody thinks only of industrialization when they think of the Ruhr. If I say I come from the Ruhr, they think I'm very dirty and that I live on a big mountain of dirt. It's very green here, and we've got nice parks, lots of forests, but I think that idea wasn't transported. Because everything was about *Industriekultur*, I'm not sure if this prejudice wasn't *greater* after that." And so, instead of telling another story rooted in the Ruhr's industrial history, Czerny explained, "we told nothing about industry, nothing about the [industrial past at Phoenix Lake]. We wanted a new story. To look to the future and not [build] another monument to the past."

The new story that Phoenix Lake was supposed to tell was of Dortmund as a normal, competitive, successful city instead of a unique region—an idea that would, it was hoped, be more legible and attractive to both locals and outsiders. It was carried out in the context of a local economic development program called the "dortmund-project" that began in 1999, by coincidence the same year IBA ended. The company Thyssenkrupp (a great-grandchild of the original Krupp company) was ending operations in Dortmund and had broken a promise to build a circuit-board factory that was to have brought several thousand new jobs to the city. In exchange, it offered to form a public-private partnership with Dortmund's Office for the Promotion of Economy and Employment and pay the cost of developing a plan to rebuild the city's economy (Jonas 2008). Thyssenkrupp hired the consulting firm McKinsey and Company, and the partnership was officially launched in 2000. McKinsey produced a ten-year economic redevelopment plan, local instead of regional, with the motto "private investment instead of public subsidy" (Jonas 2008, 9).*

Meier acknowledged that the dortmund-project "remembered the benefits of IBA" and thought of itself "not only as an economic development project [but also as a project] that must develop the urban living qualities too." But McKinsey's report and the dortmund-project were also far more explicitly, self-consciously, directly "economy-driven and employment-oriented" than IBA (Jonas 2008, 1). Though the dortmund-project was never fully executed as planned, in 2013 the city of Dortmund's promotional website still reflected an image of "the new Dortmund" that described the project as "one of the triggers for Dortmund's metamorphosis into a modern technology location" (Stadt Dortmund—Dortmund Agenteur 2015). The dortmund-project's goal was to attract new companies from what Meier described as other "B-type" locations. Though concerted efforts had produced modest economic growth in the Ruhr in the late 1990s and into the early twenty-first century, unemployment remained high, and Ruhr cities' populations declined between 2 and 9 percent from 1995 and 2007 (Bartholomae and Nam 2014, 92–93). In addition to creating jobs to retain old residents, Dortmund's economic development professionals also wanted to attract new, middle-class professionals working in new high-tech industries who might move to Dortmund with their families.

The "new Dortmund," as the name suggests, has a different very relationship to history than IBA Emscher Park did. Rather than *recycling*—figuring out how to use the past in a new way—the project inaugurated a "new local high-tech imaginary" (Jonas 2013, 2) that emphasized building new, was local rather than regional, and promoted the region through technology and quality of life rather than nature and culture. In Meier's telling, when McKinsey representatives came to Dortmund, they took one look at the town and said: "This is rubbish! Take it away! We want to make a new Dortmund!" When I suggested, incredulously, that they could not possibly have wanted to tear *all* of it down, Meier replied:

---

* Though Jonas argues that its proposed heavy reliance on city, state, federal, and EU funding meant that, in practice, the plan was in line with the region's long history of collaborative, cross-sector solutions to economic crises. He also observes that the plan was partly a result of a similar, successful economic development program McKinsey had developed for Volkswagen in Wolfsburg, though, ironically, at least part of Wolfsburg's success was due to the fact that Thyssenkrupp had built Dortmund's promised circuit-board factory there instead (Jonas 2008, 15; see also Jonas 2013).

No, they didn't want it—any of it. They said, "The Nordstadt, tear it down! We don't need more Nordstadt!" "We don't need Hörde! Your city will be like Sydney Harbor or London's East End!" They really had the idea—They said, "Take a plane, and take the city council, and we are flying to Sydney Harbor, we are flying to San Francisco—and we're flying to Toronto—and I will show you what Dortmund will be like in ten years." They said that the BIP, the net product per capita, would rise to double the amount of today. And that productivity would rise three times. They really had big ideas [laughing]. And they said, If we really can do this and make an international technology city out of Dortmund, it won't look like this! It won't look like it looks today.*

Though no project of this scope was realized, aspects of the dortmund-project's vision were internalized and "[became] part of everyday life." One idea that filtered into business as usual was its clean-slate approach to the industrial past. While IBA embraced the Ruhr's industrial history and celebrated it through the landscape aesthetics popularized at sites such as Duisburg-Nord, the city developed an antipodal presentation of nature at Phoenix Lake. I was not able to speak to the lake's designers and, thus, do not know whether the dortmund-project's desire to build from scratch directly shaped the design, but it is clear that, instead of *Industrienatur*, they built a simulacrum of first nature and that, instead of promoting the industrial and ecological history of the site, city officials deliberately designed the lake to look natural and dehistoricized. The lake reveals almost nothing about the site's industrial history or the technology required to create and maintain it. Like the portion of the renaturalized Emscher River running alongside, Phoenix Lake has soft edges

---

* Like imaginaries of nature, the dortmund-project's new high-tech imaginary was aspirational and had effects. McKinsey promised to create seventy thousand jobs in Dortmund, a number that was, according to Meier, "nonsense from the beginning" but that "everybody wanted to believe." Meier counted as one of the dortmund-project's greatest successes that it "[fed] a local consciousness of a 'we can do it!'" feeling and provided an "intellectual" as well as a "material" basis for building up a new economy. And, while he conceded, "My boss would probably . . . tell you it's based on a real economic basis," in his view "the main impact was [creating] a network and feeding consciousness." He described the project's "start-up network" of over seven hundred Dortmund businesspeople as "really a network of finding self-consciousness [of being economically strong]": "Businesspeople came together, and they confirmed: we are strong, we have the dortmund-project, and we're fighting against joblessness and so on. It really worked." He compared the dortmund-project to a "campaign" to explain how, after ten years, people have grown fatigued but nevertheless felt that the imaginary of the new Dortmund had become part of the local consciousness.

FIGURE 6.1. The "renaturalized" Emscher and Phoenix Lake, with new housing in the background. (Photograph by author.)

and artificial curves, banked dirt, bright yellow grasses, and low vegetation forming a natural barrier to the water along most of its edge (see fig. 6.1). The site contains an adventure playground that does obliquely reference its industrial history with a "nine-meter-high red forest of poles" painted yellow, orange, and red rising from the earth. But the playground is only Phoenix-*themed*; it gives visitors no indication of what kind of ash this Phoenix might have risen from (Gust 2008).

The legacy of the choice to have left so few markers of the steel plant is that it is possible for newcomers to Phoenix Lake to remain blissfully unaware of the site's industrial past. The only artifact that might properly be called *Industriekultur* is the exception that proves the rule. At the commercial end of the lake, nearest Hörde, an elevated footbridge leads to the lake's "Culture Island," originally designed as a performance space. There sits the "Thomas Pear," a seven-meter-high, four-meter-wide, sixty-eight-ton egg-shaped, metal-armored container used for pouring hot molten steel in the Thomas, or Bessemer, steelmaking process. This jolting visual reminder of the site's former life was preserved thanks to the efforts of the local volunteer-run historical association (*Heimatverein*). As the association's longtime leader explained to me, Thyssenkrupp agreed that the Thomas Pear could stay but refused to provide any financial or physical assistance. The volunteers had raised the money required to hire three cranes to move it to its current location. And though the association recommended preserving additional artifacts from the factory—in particular the one-hundred-meter-high smokestack from which Hörde's emblematic flame had once burned excess gas into the night—these efforts failed because they were, in the association leader's words, meaningful landmarks and "memory symbols" (*Erinnerungssymbol*) "for Hörde but not for the others." The fact that someone could stand at this site and have no idea that it was once a giant factory is unimaginable to anyone who has lived in the Ruhr for any length of time, so deeply is the region marked, physically and imaginatively, at every step, by industry, but the performance of nature is effective. From the shore, Phoenix Lake looks convincingly real, the Emscher biologically alive, the housing attractive, and Dortmund not unlike any number of other midsized cities in Germany and Europe.

And that was the point—using Phoenix Lake to present the "new Dortmund" as a normal city rather than a place of difference and exception. As Meier suggested, the city's goal was to attract businesses, and it pursued this through efforts organized not around hard-to-grasp abstractions such as *Industriekultur* but on the basis of a different abstraction often called *quality of life*. A brochure the city had created to promote the Phoenix redevelopment paired a photograph of the new office space at Phoenix West with a sunny rendering of boats and concessions at Phoenix Lake. The band of text separating the two read: "Zukunftsstandort für TECHNOLOGIE und LEBENSQUALITÄT" (Future location for TECHNOLOGY and QUALITY OF LIFE). If the office space represented technology, the thing that represented quality of life was nature— the green space, the woods, and, especially, the lake. These were "soft" location factors, *Standortfaktoren*, things that helped Dortmund compare favorably to other places. In Meier's words:

> When you go to the real estate fair in Munich, people come to your table, and they say, "OK, Dortmund, it's a B location, type B, but there's a lot of development going on, we've heard of the Phoenix project." . . . They've heard of the dortmund-project, but the Phoenix project really is well-known, in Germany and even in Europe, and it really transports a better image than before. The *Ruhrgebiet* really transports a very bad image, of a place that's economically weak, environmentally bad, and fun of life zero. OK? And people hear of this [Phoenix project], and they've got these images of a lake with some decent

houses around and . . . Well, it's kind of modern and a good environment, and I think this helps to make the image better.

Again, as in prior projects, nature offered itself up as an effective way to remake, in this case, Dortmund in line with new urban ideals. The city's economic development professionals thought that Phoenix Lake would help "transport an image" of Dortmund as a good place to live—a place with good quality of life—to counterbalance the negative *Ruhrgebiet* image of "economically weak, environmentally bad, and fun of life zero." While IBA used *Industrienatur* and *-kultur* to elevate the status of the Ruhr region by promoting industrial heritage sites as tourism-worthy landmarks comparable to Berlin's Brandenburg Gate or the Cologne Dome, the city administration used Phoenix Lake to position Dortmund in a competitive field of other second-tier cities. Meier was careful to say that this image change did not really "change the economic basis" of the city, but, as a marketing strategy, it did repackage Dortmund as a pleasant place and made it legible. And the city was able to do so—to envision a lake as a means of achieving these objectives, to see and use nature in this manner—even having confronted all the real imperfections and shortcomings of one of these greening projects, even having been frustrated by IBA and *Industrienatur*.

## A WIN-WIN: NATURE AS A PUBLIC GOOD

City officials were able to see the lake as a new public resource for the whole city to enjoy, regardless of their relationship to structural change or to the project itself, partly because it was serving a purpose similar to that of prior eras: providing an experience of economic decline as shared environmental gains and of structural change as improvement for all. Like the blue skies narrative taken up by *Revierparks* planners and chancellor-candidate Willy Brandt in the midst of the collapse of the industrial economy, professionals engaged in the Phoenix project argued that the lake offered good quality of life and an experience of structural change as positive, for old-timers and new residents alike. As Mayor Sierau explained:

> Being this old industrialized area, it was clear we'd need comprehensive projects in order to make clear to everybody that we are moving—changing. And of course the Phoenix project was perfect for that because, in a place where you had steelworks in earlier times, to reestablish the urban landscape, ecologically oriented landscape, that was easy as a story—to say, "This is structural change

you can realize, you can put your hands on." . . . Within ten years—less than that—we reurbanized the area. We reestablished the landscape. It was easy for people to understand. I mean, if you go to that place today and tell people, Well, in earlier times, ten years ago, twelve or fifteen years ago, there was a steel-production line here, they say, "Whoa." It's very difficult for them to imagine how it looked before, as difficult as it was for other people fifteen years ago to imagine that there would be a lake in ten or fifteen years' time. But, for those who know how it was before and how it is today, if they have that comparison, they say, "Oh, wow." It's breathtaking in a way, you know.

Sierau's description of Phoenix Lake as "structural change you can put your hands on" echoes the *Revierparks* brochure's suggestion that the parks made otherwise "incomprehensible" transformations tangible and, at least in this respect, positive. And, indeed, standing on the lake's shore on a sunny day, the air does, in fact, smell good, you can hear birdsong (assuming, of course, the construction workers are on break), and little waves lap at the banks (if there is wind).

The professionals I interviewed about Phoenix Lake—all of whom had grown up in the Ruhr with parents or grandparents who were part of the industrial economy—told the same story. They repeated the narrative of deindustrialization as a process signified by, above all, a cleaner environment and a pleasant landscape and had no trouble comparing the lake favorably with the past. Sarah Schäfer, an employee of DSW21, the utilities company that oversaw the Phoenix redevelopment as the "long arm of the city" of Dortmund, remarked with some frustration, for instance: "Everybody talks about the beautiful flame, how beautiful it was, and nobody talks about [how] we didn't have any animals in Hörde [because of the ash] or the pollution and dirt of the surroundings, and it was a bit romanticized." She was referring to the smokestack and burning flame that the *Heimatverein* had tried to preserve on the grounds of having once been important symbols of security and prosperity for the neighborhood. But Schäfer preferred Dortmund's blank-slate approach and emphasis on an unremarked-on, pleasurable experience of nature to IBA's strategy of "recycling" industrial history as a source of local pride.

Schäfer and the others I interviewed were structural change success stories. They had experienced pollution and unemployment as children, had gotten good jobs, and were now part of the "new Dortmund." They had little interest in what they saw as romanticized versions of the industrial past and no qualms about describing structural change in terms of environmental gains. When I asked, "So no one feels nostalgic?" a common answer was a firm, unhesitating no: "Most people are happy that [the industrial economy] is over," and further: "Most people are proud about how green the city is now." The landscape architect who made this last remark, who supervised park maintenance at Phoenix Lake, pivoted to describe the postindustrial Ruhr as "green" with no prompting from me. He, like the others, was comfortable describing the region's trajectory as a shift from dirty industry to a clean environment, signified above all by changes like those at Phoenix Lake.

In addition to identifying this recurrent narrative, interviews make it possible to see how this idea is discursively constructed—how people describe greening projects as well intentioned and in the public good. Phoenix Lake was marked by compromises typical of contemporary greening projects, which, as explained in chapter 5, are often eroded as public goods by cutbacks in municipal funding and declining commitments to government-provided amenities. This has meant that, like other public institutions such as libraries, universities, and museums, city parks must increasingly generate revenue, often through on-site housing or concessions. Here, Dortmund had to sell private housing to pay for the public lake. No one involved in the project had trouble describing these trade-offs: the lengths to which they went to keep the lakefront public, how as construction costs grew more and more land had to be sold off, how they "had to fight for every meter that's been made green here." But, in conversations like this, interviewees were very careful to separate the lake itself—*nature*, the public good—from the consumer goods (the housing) and the project's economic objectives. When I asked Czerny whether she thought that a project like Phoenix Lake was an effective economic development tool, whether it actually helped bring new businesses or money to Dortmund, rather than answering, she corrected me: "It [the lake] doesn't bring in money. We are very pleased if it's zero [if it breaks even]. It was possible to build the lake because of the money we won with it [we made through the housing]. And, if the project doesn't cost a lot, then it's a big success of course." She wanted to make it clear that *the lake* was not the source of the money, was separate from the project's economic agenda. She went on:

> Hörde was a suburb for really poor people because, after the factory closed, there was a lot of unemployment and really bad shops . . . a lot of bakeries and nothing else because there was no money and nobody from the other suburbs would go shopping there. And the impression was that there would be a new lake for only the rich, but this isn't the reality because it was planned [otherwise]. If you've got water, the land directly at the waterfront is often the property of private owners. So, if you build a house, you get your own [piece of]

shore[line] and your own entrance to the water. For example, at Baldener Lake in Essen, if you've got the money, you can buy [private water access]. From the beginning, it was clear that the whole shore was for the public [at Phoenix Lake]. It's public. You've got paths for walking and paths for bikes, but everyone's got direct access to the lake. And, after the direct access, after the nice public areas, after that the [private] grounds begin. And that's why everybody can use it and everybody wins. It's a win-win situation.

Czerny's description of the project as a "win-win" suggests that she believed that the site's two functions—as public good and as consumer good—*could* be separated and, importantly, that the existence of the latter did not compromise the experience of the former. Like the projects discussed in the previous chapter—from IBA Emscher Park to the High Line—Phoenix had to strike a balance between public good and private development. As with those projects, it was expected that it could serve double duty—that the sale of private experiences and amenities would subsidize a freely available public resource without detracting from it, that these spaces' income-generating functions would not compromise their status as public goods, that being surrounded by expensive houses would not change the public's experience of the lake. And, indeed, those involved in the project were insistent that the lake—conceived by a local planner, approved by the city's mayor, embraced by the city's public, and managed by the local utility—was a fiercely local project carried out with Dortmund's public in mind. As Czerny put it (her English peppered with German): "No, no. It's personal. It's only for Dortmund; it's a kind of *Entschädigung* [remuneration], a kind of *Wiedergutmachen* [making amends], for the suburb of Hörde, which was for about thirty years in a very bad situation. And of course it's a *Zugewinn* [gain] for the other suburbs and for the whole city as well—it's the most interesting leisure area here now." Schäfer also explained that the lake itself was, from the beginning, "for everyone":

It was very important that Phoenix Lake was for everyone, that it was for the masses, like Westfalen Park [a large local park] is. Because, as we've already said, the housing is rather expensive, it was really important from the beginning that the lake was for free. If you walk around it, this is a sort of leisure/nature thing that's free for everybody. That was rather important from the beginning because there was criticism of course. It's a lot of money, and people couldn't imagine how it would work or how it would look, and so that was also important for us to stress. And it *has* become that. If you go there on a Sunday, it's really—too many people, you know? It's there for everybody!

Like Schwarze-Rodrian, who described IBA's many efforts as simply "opening up" the landscape, those involved in Phoenix Lake described these machinations as efforts to provide a public good, to make nature available. Both Czerny and Schäfer made clear that the thing that could not be compromised was nature itself, and neither seemed to think that these trade-offs would affect the experience of the lake. Czerny took pains to clarify that the lake itself was not involved in the project's economic functions. For Schäfer, preserving the public shoreline was the marker of having ensured that the nature on offer was "for everyone." Even in this compromised situation, and in spite of the utilitarian purposes the site as a whole serves, both Czerny and Schäfer described the lake itself—the nature—only in terms of its intrinsic social benefits. In insisting that the lake was "for everyone"—a public good—both carved it off from the rest of the Phoenix redevelopment, separating it from the project's economic objectives. They talked about the private and for-profit components of the Phoenix project—in this case, the construction of the housing—as trade-offs that had to be made so that the public resource could be preserved.

## "EVERYBODY LOVES NATURE"

How did the public respond? As a social project, Phoenix Lake is imperfect in ways that scholars of the political economy of contemporary greening might predict. It is a public space organized around highly stratified forms of consumption. It contains no affordable housing, has insufficient public parkland, and, as we shall see, is increasingly seen as contributing to gentrification in the neighborhood. This situation is almost entirely a product of the project's relative lack of external public funding, which meant that the redevelopment had to be carried out with economic mandates at the forefront. In order to keep the lakefront public, the city had to sell off the parkland for housing. Because it had had to seek EU funding to construct office space at Phoenix West, that space stood partially empty, tied up by funding regulations. And, for want of investors, Phoenix West's blast furnace had not been opened as public *Industrienatur*. Thus, in spite of the good intentions of its creators, in 2012 Phoenix Lake was a recreational area too small for most people to use, with housing too expensive for most people to buy, and new restaurants and retail that many could not afford.

But for Phoenix Lake as a greening project—a signifier of nature—these compromises did not really matter. Among Dortmund's general public—the second receiving audience examined in this chapter—we might have expected these realities to be clearly at odds with the idea of the site as a public good.

But, as for employees of the city of Dortmund, they did not dampen public enthusiasm. Instead, the public responded by receiving greening projects as universal public goods, mirroring the intentions of the project's creators. IBA project directors' description of Emscher Park as a public good was consistent with the park as actually experienced because ample public funding ensured that equitable access and public housing could remain central to the project design. For *Revierparks* planners in the 1960s, democratic accessibility was the number one concern for the parks, as for the political moment more generally, and the spaces reflected these commitments, at least to the degree that they were large and conveniently located. Because Phoenix Lake was so visibly marked by its economic compromises, it seems reasonable to imagine that this could have produced a more measured or pragmatic account of the benefits of the project—or, in other words, that the material realities of this project would make the universal ideal harder to sustain. Certainly, I expected some discussion of the compromises that had been made. But this was not the case, not initially, and not for quite some time.

It cost €168 million to develop Phoenix Lake, not counting the cost of the land ("Dortmund 'PHOENIX Lake'" 2013), a big investment for a small city in a precarious economic situation, even a somewhat absurd one. So I asked everyone I met, "Why a lake?" Though one person referenced the historical significance of water at the site (apparently the Nazis had first imagined building a lake there in the 1930s as part of a monument to fallen heroes) and another a possible desire to return the landscape to its preindustrial state (until the late nineteenth century, the area had been swampland and part of the Emscher flood zone, a history still visible in local street names), by far the most common response, among both professionals and lay audiences, was that the city built a lake because "everybody loves water." For example, when I asked Schäfer, who had just interviewed a number of residents for a documentary she was producing about Phoenix Lake, why people were interested in buying houses there, she said:

> Everyone we ask—and I talk to a lot of people because of the documentary—they say it really is the attraction of the water. It's really the water that draws the people. I mean, we are four hundred kilometers from the sea! When you live in the Ruhr area, you can go to the Netherlands on the weekend . . . but that's about the closest you are to the water. Of course, we have the Ruhr [River], and we have some really nice areas. In Bochum, there's the Kemnader Lake, near the Ruhr University, and, if you go there on a weekend, there are really masses of people. Water draws them. And it's the same here. And everybody we talk to

says, "The water—living at the water really fascinated me, and I wanted to go there." And that's the reason for Phoenix Lake.

This simple rhetorical justification for this project—that people are just attracted to water—is a lay articulation of the idea of nature's universal benefit. People's responses suggested a tacit belief in the idea of a universal nature, that attraction to nature is a basic human experience, a commonality shared across class, ethnicity, or other forms of social difference. And the idea that "everybody loves nature" literally could not be elaborated on, despite my prodding in interviews. But more interesting than this discourse's existence is the fact that it subsequently guided practice.

Belief in nature's universal benefit affected Phoenix Lake from the project's inception by curtailing the initial planning and public decision-making process that might have preceded construction. According to Mayor Sierau, though some were initially skeptical about whether such a large project could be carried out successfully, public and professional enthusiasm for the lake ran so strong from the beginning that, once the idea was on the table, the good of replacing an apparent wasteland with a lake was never fundamentally questioned. Or as Meier put it, "No one dared question this project."

> If you knew the situation ten years ago—it was horrible! And then we had the chance to do something here. It's correct that the plans were not discussed in the beginning. Or—they were discussed, and everybody said, "Yeah, it's good!" Nobody said—nobody asked for money, for example. Even during election times, the opposition in city government didn't ask about the money, how much it would cost. Nobody dared question this project. Everybody wanted to have it. . . .
>
> [So] there wasn't really a planning process for this, a public planning process. It was said, "We want to have a lake," and everybody cried, "Yeah! We want to have a lake!" and then they made a plan, and they didn't have a real public discussion about it, everybody said, "Yeah! It's a lake, we want to have it," and so on. There were no planning competitions for this. The plan was made in the city planning department, and everybody said, "Yeah that's great."

You might want to read that quote again. Meier said that there was so much enthusiasm for the lake that the project even bypassed the city's standard planning procedures. There is a certain absurdity in the fact that a small, cash-strapped city in the middle of an economic crisis would decide to build a lake in the first place rather than investing in job training or economic development. But it is remarkable that the project was carried out without a planning competition, without a real public planning process, without sustained public engagement, and with no real challenges, not even from the opposition party in an election year.

During the time of my fieldwork, belief in the universal good of nature appeared to be shaping patterns of use too. Many in Dortmund really did love the lake, and they showed it through uses that appeared impervious to the project's distribution of benefits or the actual quality of a user's experience. Phoenix Lake's economic trade-offs constrain its use in real ways. If a common criticism of the High Line is that it has no playground and of Brooklyn Bridge Park that its private housing is located *inside* the park, at Phoenix the major sacrifice was the parkland, sold off to keep the lake public. As a result, the park as eventually built is extremely narrow—a ring of just twenty hectares of green space (Wuppertal Institut für Klima, Umwelt, Energie 2013, 83) that provides more of a buffer between the private housing and the crowds at the waterfront than an actual recreation area. As an *artificial* lake, Phoenix Lake is also a fragile, heavily managed ecosystem. It is kept separate from the Emscher River in order to prevent sediment from collecting. The water must be filtered once a year to remove phosphates and the growth of plants and seaweed curtailed. As a result, its uses are highly regulated; people cannot swim in it, dogs cannot play in it, and there are no fish. A small yacht club and boathouse were just being completed when I was there, and BMW had begun hosting an annual yacht race there, but most of the time the water was empty.

But people love Phoenix Lake just the same. One sunny Saturday afternoon in September 2012, I recorded a scene that looked, as I speculated in my field notes at the time, like Central Park must have when it was first completed. It was a hot day, and hundreds of people were walking or biking, mostly in pairs, around a crowded path cut into a bare, treeless expanse that was still part construction site. The sun beat down, the café and ice cream shop were not even open yet for a respite, and still people were determined to recreate. They were also dressed finely as though on display; my German roommate remarked that people looked like they had just come from church. The grassy areas around the lake had been seeded, but portions of these lawns were already worn away from overuse, the bare dirt the product of the park's small size and overwhelming popularity. Though such wear is a common problem in heavily used parks in large, dense cities, Phoenix Lake was the only place I ever saw it in Germany. And that day the lake was not actually even crowded! In the spring, when the retail spaces in the commercial buildings at the western end of the lake would house a popular café, a *Gelateria*, a stylish *Wurst* shop, and a pharmacy, people would be lined up eight deep in the paved area between the concessions and the lakefront, and the lines for gelato would stretch twenty feet from the open storefront, winding along the public patio area edging the lake. On weekday mornings, I often arrived to the sound of drills and jackhammers rocketing across the lake, and *still* people would be there, sitting on new benches, apparently impervious to the construction.

That day, my roommate (also new to Dortmund and from a picturesque coastal city) expressed disbelief at the scene. These people must be fools, he said, to travel to this tiny, crowded piece of artificial nature, unable even to cool their feet in the water, when real nature was not far away. At the time, I, too, was surprised, but today I would say that we should not have been. If, for greening protagonists, belief in nature as a universal good makes well-intentioned greening acts in the public good possible, for receiving audiences it provides a framework for interpreting nature experience. As I have said, audiences need not mistake Phoenix Lake, or any other greening project, for first nature in order to respond to it as a signifier of a powerful shared imaginary. Just as greening in the Ruhr challenges the idea that such projects are simply rational reactions to specific morphological conditions (density, lack of green space), the popularity of Phoenix Lake suggests that patterns of use are not outcomes of calculated assessments of the authenticity of such sites or analyses of their costs and benefits.

In 2012, as we will see below, people were just beginning to complain about the park's inadequate size and lack of active recreation areas. Use may have declined since then as the novelty of the site has worn off. And, of course, there is no way of knowing what patterns of use would have looked like had the lake been designed differently—there might have been even more users, or more diverse ones. But, throughout the several years of the lake's planning, construction, and early use, Phoenix Lake had been publicly accepted as universally desirable and beneficial and, in that sense, as nature. And, in spite of the limitations, inconveniences, and inequities built into the lake as actually designed, people were not voting with their feet to any great degree. There had been no boycott of the lake, no mass exodus to the shady, quiet paths of Phoenix West or the playgrounds and gardens of Westphalen Park, and no real breakdown of the public consensus surrounding the good of nature. In an area with many green spaces accessible by bike, foot, car, or public transit, Phoenix Lake was a clear favorite, serving as a recreational facility, leisure area, and social fishbowl for tens of thousands of curious residents.

## CRITIQUING NATURE

To say that greening projects are carried out and received as public goods does not mean that there is never any clear-eyed assessment of greening projects, never any critique or dissent. Rather, public discussion tends to take a particular form: one in which critique comes late, if at all, and in which nature's fundamental status as a universal public good is never questioned as nature is called up and mobilized by people with otherwise very different views and desires. It was not until late 2012, after the lake had been filled and two-thirds of the residential properties sold, built on, and occupied, that a sustained and coherent public critique of the project emerged. Its public face was Jürgen Evert, a photographer who had made it his personal mission to speak out against Phoenix Lake and advocate on behalf of the neighborhood's low-income, mostly Turkish families. As Meier had done with IBA, Evert rejected protagonists' claims about the project's universal benefit but still mobilized the idea that "everybody loves nature" in his arguments.

Over several years, Evert had documented a process of what he called *gentrification* (using the English word) in Hörde, which he presented in the form of a photo essay, a website, an exhibit at the local *Volkshochschule* (community college), and an evening panel discussion attended by about one hundred residents. He argued that Phoenix Lake had caused rents to rise in Hörde by increasing the desirability of the neighborhood. He said that old housing near the lake was being "upgraded" and presented photographs as evidence, such as images of front doors with empty doorbell nameplates, indicating displacement. His criticism also mirrored the contemporary scholarly critique of greening projects presented in the last chapter. In the absence of significant public funding, and in the context of a vision of Dortmund as a competitive city, the project had made concessions to private interests that had compromised the park as a public resource. And, in constructing the lake without imposing rent controls or making provisions for affordable housing, the city had produced a new green amenity that was now causing displacement in the neighborhood.

At public events that Evert organized, I heard audiences basically agree with these claims. An employee of the Emschergenossenschaft, the regional water management cooperative, for instance, was one of the speakers on Evert's gentrification panel. To great applause, she argued that equality must "be designed." She acknowledged that, as an employee of a *Genossenschaft*, or a cooperative, she did not have the same mandate to balance economic growth and social responsibility as the city did and was a bit more free to keep the public interest in mind. Still, she thought that Dortmund's mistake had been to pit private interests against public quality of life in the first place. During the question-and-answer period, some asked whether displacement in Hörde could properly be called *gentrification* because there were such high vacancy rates in comparably priced housing in the city overall, some affirmed that there was a real need for middle-class housing in Dortmund, and some wondered whether the rent increases or the numbers of the displaced were as great as Evert suggested. But most seemed to accept his critique. Even Meier agreed that "what you *can* complain about is that there [was] no consequent scheme of publicly funded housing for people in this area," even though he took issue with some of Evert's other claims.*

In this way, the public's criticism of Phoenix Lake repeated a pattern visible in Dortmund's criticism of IBA. In these moments of discussion focused on housing and broader neighborhood change, it was possible to assess Phoenix Lake as an economic development project rather than a nature/public good project that no one dared critique. Evert and others accused the city of having been utopian in its lack of social planning and offered planners seduced by nature a critique of Phoenix Lake as an economic development project with rippling social and economic effects. In both cases, it was receiving audiences confronted with the day-to-day realities of these projects who reminded greening protagonists that these were not simply gifts of public goods but more complicated sites affected by social interests and political economy.

But the discussion also highlighted another interesting dynamic accompanying nature as a matter of public debate, which is that the good of the lake was never questioned—that, when critique arose, it was not a critique of *nature*. Even the most vocal critics of the Phoenix project separated the lake (preserved as a public good) from the project as a whole. Schäfer's and Czerny's strong belief in the idea that "everybody loves nature" and in the lake as a

---

* The tenor of the discussion was, of course, shaped by the German context, especially the country's strong public sector, a general belief that it is the role of government to provide public services, and much less faith in equitable access as a natural outcome of market mechanisms in comparison to opinion in the United States. As was repeatedly pointed out to me, Germans hold the public sector responsible for quality-of-life improvements far more than Americans do and expect government to invest in and maintain public green spaces without relying on the volunteer labor or public-private partnerships typical of the United States, just as it provides health care, housing, and pensions. Echoing the Emschergenossenschaft employee's words, the shared assumption is that an equitable society must "be designed" and that this is the public sector's responsibility. In the case of Phoenix, this should have come in the form of the missing "social plan" that typically accompanies such economic development projects. Though Meier maintained that economic development in Hörde did serve the public good—that the parts of Hörde in question were so troubled that there was no real danger of gentrification per se and that the lake would help everyone by catalyzing investment in infrastructure, schools, retail, and public services.

public good meant that, regardless of the compromises that had to be made around the lake (such as the housing), they believed that the lake's status as a universal public good (as nature) would not be compromised as long as the waterfront remained public. Similarly, I never heard Evert or anyone else reject or even fundamentally question nature as a public good—by claiming, say, that the lake would be better understood as an economic development project than a park or by arguing against investments in nature in favor of jobs or housing. No one argued that it was impossible to ask nature to serve as a public good and a means to private investment simultaneously. Instead, nature was claimed and reclaimed in the same terms: as an objectively available, universally beneficial public good that all people desired, benefited from, and deserved to have access to.

Evert, for instance, brought his critique of Phoenix Lake home by invoking nature. He counterposed new residents' gain in access to nature with poor people's loss, describing Hörde's Turkish residents as having lost their *Pantoffelgrün*, or "house-slipper green space" (i.e., the kind of green space one could access in one's house shoes), in their relocation. He showed photographs of fruit trees and small backyard or rooftop gardens being destroyed. He described those forced to relocate to affordable housing elsewhere in Dortmund as *gärtenlos* (gardenless). The injustice as he saw it was that, after having suffered through the pollution of the factory, the mess of its removal, and the noise of lake construction, Hörde's poorest residents would not now benefit from the new nature amenity.

And with Evert's invocation of urbanized nature came the social dynamics we have come to expect: some whisper of paternalism. When Evert took it upon himself to speak for Hörde's relatively marginalized, disempowered public, he shifted from critic to advocate, from receiving audience to possible greening protagonist, much as Roland Günter had done for Eisenheim's residents fifty years before. And sympathetic audiences took up this discourse, arguing that one of the public sector's primary roles is to "care for" its citizens, intervening in order to create a more just society for all. One audience member went so far as to compare the city's proper role to that of the Ruhr's nineteenth-century industrial barons, saying that, though they were exploitative, Krupp and others had taken better care of their employees than the city had of its citizens at Phoenix Lake. The critics of Phoenix Lake are neither a self-identified nor a formally organized group. But they bear more than a passing similarity to other actors described in this book in that they possess an aspirational vision of a better social future, genuine good intentions, relative power and influence, and a benevolent relationship to a population of interest.

*Experiencing Nature as a Public Good* 197

FIGURE 6.2. Turkish users of Phoenix Lake. (Photograph by author.)

In addition, like all the greening protagonists discussed in this book, critics also took on the role of speaking for a target audience that was real but abstract and not part of these conversations. There was consensus that Phoenix Lake did little to help—and more likely harmed—the neighborhood's two most vulnerable demographic groups, its traditional blue-collar working class and its Turkish immigrants. At the time, Hörde had about a 20 percent unemployment rate, and 40 percent of residents were first- or second-generation immigrants, rates higher than those for the city as a whole, which were 16.5 and 30 percent, respectively (Stadt Dortmund 2012a, 7; Stadt Dortmund 2012b, 11). But, like the wealthy inmovers, these populations remained abstractions in these discussions. No members of either group were visibly present at the gentrification discussion or exhibit opening; both became instead new versions of IBA's imaginary friends.

The care that Evert and others took to represent the Turkish community on whose behalf they were speaking highlights to the importance of these groups' presence in critics' imaginations. I took the photograph shown in figure 6.2 at Phoenix Lake one afternoon while watching three generations of women sit on a bench overlooking the lake. I was intrigued by the first evidence I had seen of Hörde's Turkish population at the site, but more interesting than the photograph itself were the reactions it provoked.

After I shared it in a presentation in my host department at the local university, a colleague wrote to ask where I had found the picture and whether she could use it in her own writing and presentations about Phoenix because it so clearly spoke to the issues of class and ethnicity surrounding the debate about the "new" and "old" Hörde and their relationship to the lake. Evert's arguments also focused on the physical displacement of Hörde's Turkish residents rather than on, for example, former factory workers who might feel alienated by the changes in the neighborhood. The special interest in the Turkish community was partly because this group suffered from above-average poverty and discrimination in Dortmund, but also because this community's presence at sites like Phoenix Lake was a powerful image of the kind of democratic access and universal benefit that greening protagonists had envisioned. Recall the *Revierparks* brochure's picture of Turkish immigrants playing cards. Similarly, the cover image chosen for an edited book about IBA Emscher Park offered a glimpse of an IBA art installation in the background but featured a woman with a glowing red headscarf, in sharp focus and color enhanced, as the center of the frame (Reicher, Niemann, and Uttke 2011).

The central imagined threat—the new middle-class professionals building expensive homes at Phoenix Lake—may not have actually existed either. The gentrification debate was driven by the assumption that new, wealthy people were displacing the old Hörde. During the time of my research, the city had funded programming, outreach, and beautification efforts in the neighborhood in response to worries about the old Hörde being perceived as forgotten by the city, efforts premised on an understanding of the new residents as outsiders who had to be integrated—Austrian tech executives whose offices had relocated to Phoenix West, enticed by the lake and satisfied with the neighborhood's newly renovated metro station, grocery store, and sidewalk cafés. Critics shared this imaginary. One member of the gentrification panel argued that Phoenix had created a dramatic meeting of the "old" and the "new" Dortmund, a strange marriage of winners and losers in the game of structural change. He, like many others, worried about a situation of cultural alienation in which longtime residents of the old Hörde failed to integrate with the new or be served by its amenities.

But, as with the failure to attract new businesses from outside the area, it appeared that the majority of the properties were actually purchased by locals. Czerny said that most of the people buying property at the lake were actually from Dortmund or nearby towns just outside the Ruhr. "There is nobody from Cologne, for example," she said, "who has started to work here and lives here now." She described lake residents as young couples with children buying their first home or older couples who left suburban houses with gardens in order to "live more urban lives." Schäfer speculated that the young families had probably used inherited savings to start their own homes, thanks to frugal parents and grandparents who had survived wartime scarcity. A few local celebrities did live at Phoenix Lake, such as players for the city's world-class soccer team, Borussia Dortmund. But these figures were close to public property themselves—hardly strangers. And, though Schäfer conceded that it was probably not "simply blue-collar people" buying houses at Phoenix Lake, she emphasized that for the most part "it's not as posh as [she] would have thought either" and certainly "not the middle-aged doctor couple you might imagine."

## THE NATURE OF NATURE AS A PUBLIC GOOD

Aside from its scale—a large project in a small city—Phoenix Lake is typical of greening projects under neoliberalism in terms of its financing, politics, and outcomes. But this chapter documents a pattern that transcends the current political economy: that shared beliefs in nature's universal benefit condition greening projects to be carried out and received not as normative judgments about urban society but as investments in the public good. Perhaps surprisingly, this dynamic frequently coexists with critiques of nature projects, even within one individual. Individuals are perfectly capable of experiencing greening projects as social projects. Everyone I interviewed could detail their shortcomings, imperfections, and unintended consequences. But this awareness did not prevent the same individuals from mobilizing urbanized nature, either as an enthusiastic user of or as an advocate for other greening projects. In other words, urbanized nature remains popular in spite of critique. The effects of this imaginative power are felt in politics and public life at several levels.

First, nature's discursive power shapes public debate. In spite of the relatively short history of greening as a social practice, in many places the value of parks and green space as public goods is doxic. The idea that nature improves the social literally "goes without saying because it comes without saying"; the greener-equals-better formula has been naturalized as everyday common sense (Bourdieu 1977, 167). In this way, urbanized nature serves an enabling function by legitimating acts made through it. At the same time, nature's moral authority also acts as a limiting function by foreclosing discussion and keeping some topics off the table. Again, to note the dominance of these patterns is not to suggest that no professionals or members of the public ever come out against greening projects or that people are unaware of such projects' problems, but, to the extent that greening projects are treated as universal public goods, they tend to "demarcate the bounds of permissible discourse" (Lears 1985, 568, 569) by making it very difficult to be against them. In the case of Phoenix, the idea of a lake was so immediately popular that people forgot to have a discussion about possible alternatives, and it became politically dangerous to be against it. Because the lake was understood to be obviously in the public interest, and because there was broad public enthusiasm for the project, it became difficult to treat it as an economic development project with social consequences that should be examined as such. In short, the hegemony of urbanized nature means that, to be against greening projects, you must position yourself against both the ostensible common good and everyday common sense.

Second, the self-evidence of nature's value shapes material outcomes. In the case of Phoenix Lake, the power of urbanized nature meant that options other than a lake were not seriously considered and that the design itself was not subject to a full public debate or a competitive planning process. It was precisely because support for the project was so strong—because "no one dared question this project"—that no design competition was held and the planning process was rushed. Had it been treated as a "social" project—comparable to the construction of a mall, a parking lot, or a new housing complex—rather than a "nature" project, it almost certainly would have undergone a lengthier and likely contested process of planning, approvals, and public input prior to construction.

Third, belief in the good of nature allows well-intentioned action. Urbanized nature makes it possible for people to act in the name of the public good, which in turn primes greening projects to be well received in spite of their inevitable shortcomings and imperfections. As a result of this imaginary, nature's fundamental good is never questioned, making it possible to sustain a belief in the basic experience of nature as pleasurable and beneficial even in the face of highly compromised projects and materially unequal outcomes. Those who worked on Phoenix Lake were adamant about the public benefit of their efforts and of the lake itself even as they acknowledged problems with the housing. And even the project's harshest critics offered no critiques of nature; they simply wanted better access for Hörde's Turkish immigrants. In short, greening becomes a very powerful way of organizing social life because nature's normative associations make projects that involve it appear to stand outside the terrain of civic negotation. Urbanized nature conditions greening protagonists to view themselves as benevolent actors providing public goods, rather than as managers constructing ideal cities and citizens, and audiences to receive the projects as desirable public goods to be evaluated, perhaps, in terms of their accessibility, amenities, or distribution but not whether or not they are "good" at all.

CONCLUSION

# Global Greening Today

This book has offered a new explanation of the origin, spread, and politics of modern greening—as a social practice that is a product of a social imaginary—in order to provide a new perspective on an activity we think we know well and denaturalize urban greening as a global, contemporary phenomenon. It is an account that lies at the intersection of two long-standing binaries—nature and the city, and materialism and culture—and one that centers a recurrent behavior whose commonalities across time and material effects on urban space and public life have frequently been lost in the cracks between disciplinary fields. It has established a view of greening as a mode of moral action, rather than reactionary antimodernism, the knee-jerk response of disaffected urbanites, or the simple provision of public goods—a mode through which people have regularly intervened to "improve" cities as physical and social environments for 150 years.

This explanation also brings urbanization into focus as a process of social transformation that produces new forms of everyday life and social consciousness as well as physical and economic changes, and emphasizes the effects of this social imaginary on urban practice and, thus, urban change. What did urbanization "do" to nature? It helped produce and contributed to the spread of a social imaginary of green as good by turning everyday signifiers of domestic nature from direct, material goods valued for subsistence purposes into indirect, moral goods seen to be good for you. As greening interventions are used aspirationally, to represent new visions of society and to help create them, these practices have had very real consequences. They do not simply produce decorative, quality-of-life improvements, but materially embed a normative vision of society in the landscape and physically reconfigure how people live and inhabit it. In the Ruhr, greening has remade the relationship between work and home, created a legible urban form, and supported the formation of bourgeois publics for 150 years. Beyond the Ruhr, it has transformed landscape and settlement patterns, and has shaped social life by designating who encounters whom where, what activities are made public or private or intended for individual or collective participation, and what kinds of spaces are created for politics.

In documenting the effects of urbanized nature on the transformation of urban environments, this argument also underscores that social imaginaries are not just social facts; they are realized materially. As Andreas Glaeser has put it, "discursive understandings of society" must be "laminated" onto "kinesthetic ones" (Glaeser 2011, 175). As Bob Jessop writes, "imaginaries provide not only a semiotic frame for *construing* the world but also contribut[e] to its *construction*" (Jessop 2010, 342). Nature's role in material engineering has classically been shown in efforts to establish the nation and, in a sense, greening as citymaking is the same activity at a different scale. But this book is particularly concerned with the conditions of possibility for using nature as a medium and mechanism of control as well as a site of control. Human societies have acted on nature in many ways—preserving "wildernesses," extracting natural resources, taming bodies with culture and technology—but they have also created greening as a means through which to construct ideal social worlds. As historical work on nature and nation suggests, this activity is also a spatially and temporally specific one. Greening contributes to producing a particular scale and form of social organization—cities— within particular parameters: those of urban society.

This is an account that builds on disciplinary insights in both sociology and urban studies. Sociologists are intermittently sensitive to the "materiality" of culture (Griswold, Mangione, and McDonnell 2013; see also McDonnell 2010; and Rubio 2014) and view elements of the built environment—from buildings (Gieryn 2002) to everyday objects (Molotch 2003)—as artifacts of the social, reflecting historically specific views of the world. This is especially true of scholarship on the social construction of nature, which has shown how things such as pigeons, gardens, landscapes, and natural disasters reflect particular social worlds and visions of common culture (Jerolmack 2007), class (Bell 1995), and nation (Fourcade 2011). For urbanists, meanwhile, it is more or less intuitive to perceive design interventions as social engineering of this kind—for instance, to view Haussmann's Paris boulevards and Robert Moses's New York City expressways as regimes of technosocial control (Harvey 2006;

Winner 1980). Yet, while the design of urban green space has long been taken to be central to the physical and social life of cities, urban histories can easily overlook nature or relegate it to a chapter on parks rather than systematically examining, for example, green space as an organizing principle for urban publicness. And, at least in the popular imagination, fleshy civic transformations are still more easily traced through changes in the design of streets, plazas, and buildings. Nature is less often understood as a tool used to *realize* those visions—as a material deployed in the realm of the civic, at the scale of the city—certainly by those wielding it.

This study is intended to demonstrate the power of such a view for understanding landscape transformation and the politics of the built environment in a superficially atypical case, a site that is at odds with commonsense understandings of the practice rooted, as they have been, in large cities at landmark moments of urbanism. The Ruhr's unusual urban form and long history of greening test these explanations' limit conditions and also more closely resemble the conditions under which the majority of greening has taken place, historically and in the present: in places where greening is not an obvious or the only response to social problems, that are not suffering from crushing density, that have other imaginative frameworks for understanding nature and improving society available, and that often lack a large urban bourgeoisie. So what resources does the Ruhr offer for understanding greening's widespread uses and effects today?

## TWENTY-FIRST-CENTURY GREEN CITIES

First, this book is an argument for seeing the contemporary interest in green cities as continuous with greening's urban history. Today, greening—carried out in the name of sustainability, ecomodernization, climate adaptation, and quality of life—remains central to both development-oriented and grassroots projects in classic industrial cities such as New York and London as well as to community revitalization and economic and social renewal initiatives in suburbs (Keil 2017) and "shrinking" cities in Europe and North America (Haase 2008; Schilling and Logan 2008; Tumber 2011). The green-as-good logic is at work in "world class" citymaking (Patel 2015), in new, "smart" ecocities (Anguluri and Narayanan 2017), and in everyday adaptation efforts (Govindarajulu 2014) in Asia and the Middle East. Even the United Arab Emirates—the country with the world's highest carbon footprint per capita in 2006—is transforming Dubai into an "improbable green city" of farmers markets and yoga classes, golf courses, biodomes, ski slopes, and artificial lakes (Kunzig 2017). All these are places that, as the Ruhr did at the end of the nineteenth century, lack the morphology, local climate, population, and/or social history that would require or even recommend greening to take these forms or look these ways.

Aspects of these urban-ecological challenges and technologies are distinctly contemporary. When I started writing this book—certainly when I started thinking about this book—the very idea of a sustainable city was still considered something of an oxymoron among urban professionals (Angelo and Wachsmuth 2020). At the end of the twentieth century, nature's role in cities was most frequently understood in terms of economic development and livability—parks and farmers markets as a favorite habitat of the creative class—and most ecological concerns were related to the remediation and/or reuse of industrial infrastructure, from polluted brownfields to working waterfronts to rail lines like the High Line to Emscher Park. Those concerns seem almost quaint in 2020, as New York City counts on one hundred acres of live oyster reef as a buffer against climate change–induced storm surges and Ethiopia contends with soil erosion, flooding, and deforestation by planting 350 million trees in twelve hours. Today, the social goods that nature is understood to deliver are related to twenty-first-century concerns. New, high-tech forms of greening include "green building" to increase energy efficiency, "green infrastructure" that is more resilient in the face of storms, flooding, and other threats, and "green planning" around issues such as food acquisition, residential density, and public transportation to reduce carbon footprints and greenhouse gas emissions.

Nevertheless, this book suggests that, in spite of twenty-first century aesthetic and technological forms and concerns, the current interest in sustainable cities is the most recent iteration of fundamentally the same practice. Each of the three periods examined in this book exhibited a distinct greening paradigm—industrial/cosmopolitan greening, deindustrial/democratic greening, and neoliberal/consumer greening—in which projects' governing ideologies and practical execution reflected the needs and political economy of their era. While the specific concerns and technologies of the present paradigm may be new, its form is familiar and congruent with the past. The new ideals of cities and citizens are visions of "resilience" in the face of environmental threats and of sustainable urbanism as offering the possibility of a new harmony between city and nature. The protagonists are international networks of consultants and foundations working in partnership with city leaders to develop and implement global best practices as international agreements fail. And the notion of the public good reveals a growing concern for equity on the part of planners, the public, and elected officials, in response to the growing awareness that the world's poorest are most vulnerable to climate change threats globally, and to environmental displacement within affluent cities.

Much contemporary discourse suggests the opposite. Scholars and practitioners argue that climate change can or must bring about a decisive break with the past as illusions of human domination over nature become ever more untenable. Public discourse and planning practice reassert the rights of nature and call for less androcentric forms of urbanism, from "rewilding" efforts, to daylighting urban rivers, to planning cities to adapt to storm surges or sea-level rise. As Western economic and political hegemonies are challenged in global geopolitics, activists might hope or expect to see non-Western forms of nature knowledge rising in prominence and legibility in urban planning and activism. But, even as contemporary scholarship and practice has, undoubtedly, challenged inherited conceptions of city/nature and society/nature relationships, this book argues that the representations, imaginaries, and practices of greening that are dominant today are direct legacies of those nineteenth-century ones and, as such, likely creating similar patterns, blind spots, and problems.

## TRAVELING GREEN URBAN IMAGINARIES

The historical explanation for greening in the Ruhr is also useful for explaining why greening is so ubiquitous—and aesthetically similar—across such a wide range of very different environments. Traditional explanations for greening as a reaction to the conditions of the industrial metropolis can no more account for urbanized nature's travel to the Middle East in the twenty-first century than they could its travel to the Ruhr at the beginning of the twentieth because, in both cases, global urban processes, not local urban form, explain its appearance.

Greening in the Ruhr was surprising because of the region's physical conditions, not because of its particular cultural or historical context; the European and North American conceptions of society/nature relationships that urbanized nature reflects were perfectly legible in Germany. But it is both historically and theoretically relevant that early twentieth-century protagonists were able to "green" the Ruhr because it demonstrates that, although urbanized nature may have initially arisen in industrial cities, its legibility has long exceeded those places. Greening in the Ruhr was enabled not by local morphological conditions but by transformations of urbanization that were global in scale, even as they played out differently and at different times in specific places. As a result, the timing of both the emergence of urbanized nature and self-conscious efforts to remake cities by greening them is historically variable. In

France, for instance, the French Revolution marked the beginning of the transfer of royal gardens and woodlands to the public, followed by the construction of public parks in the nineteenth century (Ives 2018), while the first efforts to establish public/private distinctions through green space did not occur until the beginning of the twentieth century in the Ruhr. When they did, it was because industrial urbanization had made greening possible in two ways. At the level of individuals, it created opportunities for experiences of nature as a socially beneficial leisure space. Though perhaps initially concentrated in large industrial cities, just as experiences of nature as leisure transcended such cities, the legibility of this idea did too. At a societal level, industrial urbanization also expanded and transformed the global networks of material and social exchange—of labor, goods, and ideas—through which this idea of nature traveled. In American factory towns in the nineteenth century, as in the Ruhr at the beginning of the twentieth, elites comparing themselves to industrial cities brought urbanized nature home, regardless of local morphology or local inhabitants' cultural reference points.

The argument of this book is that contemporary greening is best understood as the global spread of this historically Euro-American imaginary: as urbanized nature traveled to the Ruhr at the end of the nineteenth century, it has traveled to Asia and the Middle East today. This is not to say that urbanized nature is or ever was the only imaginary available or that its diffusion was linear or flowed only in one direction, or to make any claims about the historical emergence of greening in any particular location, but to insist that such similar global manifestations are not solely outcomes of local conditions. There were and are, of course, many other visions of nature, of urbanism, and of society/nature relationships available in the Middle East, just as there have always been other ideas of nature available in Germany. Very different conceptions of antiurban and antimodern nature were mobilized differently by the Third Reich and, later, the Soviets in East Germany (Brüggemeier, Cioc, and Zeller 2005). And, though tracing this history is beyond the scope of this book, urbanized nature and nineteenth-century city ideals were themselves shaped by earlier global colonial encounters (Grove 1996). But the argument made here is that a specific social imaginary that gained hegemonic status through its presence in industrial cities in the nineteenth century maintains that status today. Urbanized nature continues to circulate through global best practices and is replicated by international elites as well as grassroots movements, in the same materially interconnected system within which shared logics have made common imaginaries possible, taking different concrete forms in different places as it interacts with local contexts. And, in much the same way that greening became possible in the Ruhr once nature became a recognizable way to better the social world, greening is possible in twenty-first-century Abu Dhabi because it is part of a widely legible repertoire of social action. The production of spectacular displays of nature completely foreign to the desert suggests that the desired "greenness" of desert cities has less to do with ecological outcomes or other local conditions than the fact that such displays are a recognizable part of the performance of global urbanism, a demonstration of technological prowess, and a representation of good quality of life.

Again, the analogy with nationalism is instructive. The idea of nation, too, was historically a European concept. There is no particular reason why we should live in a world of nations—why people and continents with very different histories and forms of social organization should adopt this understanding, should come to organize and manage their territories and people in this way. But, as Anderson argued, the idea of nation "spread" through markets, media, and domestic and international conflict. Embedded in a common global system, concepts traveled—were enforced or imported—along with people, materials, and capital. These were the conditions of possibility for a recognizable package of naturalized beliefs about the nation to be taken up in different places and in novel ways. And this imaginary was not available only to decision makers at the top or to those in direct contact with international actors. It became an idea mobilized by individuals in a range of ways and places, and reproduced through everyday practices—raising the flag, saying the pledge of allegiance, participating in revolutions, supporting national sports teams—as well as through large-scale, top-down interventions.

The global presence of green urban initiatives and aesthetics is often described as a policy transfer (Montero 2017; Peck and Theodore 2015). And global greening is, in part, a product of traveling best practices. But greening must be understood in the context of the broader social and material conditions that produced those practices in the first place. It is a policy transfer predicated on a shared social imaginary—a shared categorical framework for interpreting and constructing the world—that was itself produced by industrial urbanization, a process that also contributed to creating the networks on which these practices continue to travel. In Castoriadis's words, certain "relations between individuals and groups, behaviour, motivations are not simply *incomprehensible* for us, they are *impossible in themselves* outside of [a given] imaginary" (1997, 161). In my words, urbanized nature—a shared understanding of everyday signifiers of nature as indirect, aspirational, and universal goods—actually makes greening possible. Without these associations, one could not give a park as a "gift" or see urban agriculture as "transformative," and efforts to achieve economic objectives, or global city status, or to provide these perceived benefits through greening projects would be meaningless.

## PUBLIC GOODS AND THE FAILURES OF UNIVERSAL ASPIRATIONS

This perspective on greening emphasizes its similarities to imaginaries and practices surrounding other public goods by design. That greening is not an escape from society but a way of constructing society, that greening efforts are often carried out unreflexively and with good intentions, is much like the provision of nation or tradition, art, religion, or education. People leading efforts to provide schools, houses of worship, or public art are likely to see themselves not as advocating for particular visions of society but as providing desirable amenities. As with the idea of *nature* in the case of greening, associated, affectively loaded concepts such as *peace*, *culture*, or *community* help maintain an understanding of these projects as benevolent gifts rather than partial outcomes of political and economic negotiations. And like greening, in spite of protagonists' best efforts, schools, churches, and public art are likely to consolidate cultural biases in the relationships with the environment they lay out and amplify normative notions of society.

If greening projects have universal ambitions but cannot help but be partial (social) accomplishments, it should be no surprise that they will reflect existing inequalities in the landscape and the social dynamics of the world in which they were produced, in terms of how and by whom they are carried out as well as the biases the spaces themselves ultimately reflect. To use this book as an example, it will have been hard to miss that I have told a story of white men greening. This is in part a reflection of the fact that this book is a study of greening as citymaking, involving large-scale projects with a broad civic vision and the ambition to speak on behalf on an entire imagined public. The leaders of such projects will be those with the authority and the resources to green on behalf of an imagined urban public and, often, those active in networks or institutions with the resources to carry them out. Even in the counterhegemonic case of Eisenheim, Günter and his university collaborators—those with some status and influence as well as fluency in the discourse of the bourgeois public sphere—were central to the working-class movement's success. This situation reflects the fact that greening projects fall short of universality not only in their normative content but also in the demographics of their main protagonists.

But though lay publics and urban scholars know that public goods are always imperfect—the notion is invoked in unequal environments and produces unequal distributions of sacrifices and benefits (access to schools and cultural institutions, for example, is highly stratified by race, class, gender, and ethnicity)—when these shortcomings come to light, as they often eventually do in the case of green space, we still tend to treat them as historical anachronisms or local or individual failures to be overcome rather than as dynamics intrinsic to greening, just as they are to the provision of any public good. Take Olmsted's Central Park, perhaps the classic example of a green public space fallen short of its universal aspirations. Its definitive social history, *The Park and the People*, provides a critical history in terms of contested meanings of publicness: the displacement of the poor and marginalized by a small group of "gentlemen advocates who claimed to represent the entire 'public'" but who ended up producing a bourgeois cultural space in an Old World/European model, an "elite park" that allowed no sport or fun and that people had to be "trained" to use (Rosenzweig and Blackmar 1992, 7–8). In his masterful history of urban nature in New York City, Matthew Gandy asks: "What kind of public space was envisaged by the construction of Central Park? Can we really conceive of Olmsted's vision as universal?" He describes Olmsted's vision of a park for a heterogeneous public audience as being inspired by ideas of "republican inclusiveness in urban design" and as part of the "articulation of a nature aesthetic for a wide public audience" in nineteenth-century America, but he draws much the same conclusion as Rosenzweig and Blackmar do, arguing: "Olmsted's vision was never a public one from its inception but that of an urban elite who successfully imposed it on the wider society" (Gandy 2002, 99, 98).

This book suggests that asking whether Olmsted's ambitions could really have been universal because the park he produced was elitist is the wrong way to think about these questions. Instead, both these things are—can be, have to be—true. It is quite possible that, like the greening protagonists examined in this book, Olmsted believed in the park's universal benefit and was acting on a benevolent and well-intentioned impulse: to create a park for a universal public, however he imagined it. It is equally possible to see how that universal impulse remained unfulfilled in the park's actual design and management, which could not help but reflect Olmsted's own understandings of what a good city and good citizens should look like as well as expectations for use and maintenance that were biased toward the park's wealthy, carriage-owning neighbors. *The Park and the People* ends optimistically, having traced a trajectory of democratization across the park's 150-year history, suggesting that perhaps things are better today. But today—as Gandy's analysis also shows—Central Park remains in many ways an elite project: managed by a nonprofit conservancy funded in part by wealthy neighbors, still speaking on behalf of an imagined public, better resourced and better maintained than its Harlem counterparts, and still reflecting distinctly Euro-American conventions for uses of public space in its design even as it is simultaneously "used to illustrate an abstract, normative ideal of an inclusive public sphere that is held to have existed in the past" (Gandy 2002, 105).

To say that parks are always less public than they appear is not a novel statement—really a restatement of the fact that "nature is social" in spite of its appearances. But, for all our academic understanding of green spaces as human-made, when talking about providing universal access to nature—not as a normative ideal but as a real possibility—practitioners and scholars alike are still prone to act as though greening projects can somehow transcend the limitations of their time and place and the subject position of their protagonists. Certainly, Central Park is more democratic than it was a century ago, and, certainly, it was unequal and elitist in ways that were unique to the era. But Central Park is not an exception—it is the rule. Olmsted was not a deficient *individual*, a social reformer unable to get a handle on the narrowness of his own bourgeois beliefs and design a park that could truly serve the public. Instead, as a historical figure he embodies the paradox intrinsic to greening—this combination of genuine universal aspirations and normative and practical limitations. Rather than turn these into questions of historical elitism or true intentions, this book suggests that greening projects will, like any kind of social project, always reflect the world in which they are produced, often in spite of the explicit intentions of their creators, and should be treated accordingly.

## THE PHENOMENOLOGY OF NATURE EXPERIENCE

Urbanized nature does, however, have some qualities as a public good that make it unlike culture or nation or tradition, most relating to specific associations that hang on conceptions of nature and to the material qualities of these signifiers. This book is not about the word *nature* as it appears in discourse, though it uses the term for convenience. Rather, it is about physical representations of nature in the built environment. Though signifiers of everyday, domestic nature are potentially highly variable, only a limited number are called on to represent ideal societies and notions of the good. These signifiers, familiar to us from urban parks and petting zoos, include trees, flowers, grass, and green open space and, less often, small domestic animals such as rabbits, chickens, and racing pigeons. Many other forms of urban and nonurban nature—climate chaos, weeds, empty lots—are not understood to be *green* in the normative sense, as evidenced by the fact that empty lots, seen as blighted or dangerous, can still be described as having been *greened* if transformed into community gardens.

The most distinctive quality of urbanized nature as a public good derives from the phenomenological pleasure and psychophysical benefits that trees, flowers, and open space are understood to provide. When I speak about this research publicly, someone usually mentions that, as healers, philosophers, and spiritual leaders from Eastern and Western traditions have been aware for at least a thousand years (Ulrich 1993), exposure to nature has real, universal health and psychological benefits: that hospital patients with a view of trees or gardens outside their window heal, on average, a day faster and need less medication for pain or anxiety (Franklin 2012; Ulrich 2002), that a walk in nature (known in Japan as *forest bathing*) lowers blood pressure and cortisol levels (Park, Tsunetsugu, Kasetani, Kagawa, and Miyazaki 2010), and that dog owners live longer (Mubanga et al. 2017).

And, of course, all these things are true. In his classic essay "The Trouble with Wilderness," the historian William Cronon is careful to point out that, in describing the wilderness as a human idea, he does not mean that it is "merely our own invention"—that the sound of a waterfall, its cold mist, the sharp smell of wet stones, are not *real* or really pleasant; rather, he means that "what brought each of us to the places where such memories became possible is entirely a cultural invention" (Cronon 1995, 70). The point here is a similar one. At a biophysical and psychological level, nature really does clean the air, make us heal faster, and lower our blood pressure. To call urbanized nature a social imaginary does not mean that it is a false belief or that nature's imagined social benefits are simply *imaginary*, that is, made up or invented. It is not an argument against the pleasure and health benefits of pets or urban parks or windows in hospitals and office buildings.

Rather, green becomes a powerful vehicle for social projects because of these qualities and associations. The sensory conventions and forms of aesthetic practice that surround nature are different than other things understood to be good for you, such as art—even in similar institutional settings (Mangione 2016). Nature's physical qualities and appearance as unmediated give greening projects a perceived universality of access and experience that is unusual. Unlike other social engineering technologies, such as roads and infrastructure or even schools and museums, greening projects are prone to look natural and, therefore, not like tools for management. The erasure of the social is constitutive of experiences of nature projects in a way that is simply not true of education or public art projects.

Nature's tendency to obscure—a frequent target of critical scholarship—is one effect of this erasure. Its appearance as separate from the human world and timeless and unchanging in contrast to it conditions views of apparently natural landscapes as devoid of human histories and drives efforts to restore them to a mythic presocial state. Cronon's wilderness essay recounts the role of Europeans' perceptions of the American landscape in justifying the violent removal of Native Americans. The geographer Shiloh Krupar's (2013) work on the Rocky Flats National Wildlife Refuge has shown how the illusion of "pure" nature erased the "radioactive legacies of the Cold War" at a nuclear weapons plant turned wildlife refuge. The greenwashing of polluting corporations and products is another example of the power of nature to recast something as positive (healthy, good for you, good for the environment) so that you might even pay more for it.

But the experiential qualities of nature's material referents affect greening projects' political dynamics and material outcomes in other ways as well, by making nature especially available for remaking the environment. The discursive and behavioral dynamics that this research has documented suggest that the powerful associations with nature also allow greening to be carried out with good intentions and to fulfill real desires. People are not naive givers or receivers of nature. At some level, we know greening projects are social projects and are not prone to confuse these sites with "real" nature, but nevertheless we can and do mobilize this imaginary and enjoy the sensory experience that a park or an artificial lake provides. In other words, greening is not only the imposition of paternalistic desires; it is also a good desired by publics, even if never perfectly executed and never actually universally beneficial. We all inhabit landscapes, all the time, that reflect the desires of those other than ourselves. All kinds of choices—project financing, housing, the inclusion of playgrounds or sports fields, seating areas or public bathrooms, the price of concessions, to name just a few that are common in parks—reflect the normative orientation of funders and designers and their assumptions about the race, class, gender, and ethnicity of users. While these certainly affect the uses and users of green space, they do not make these spaces less real, less desirable, or even necessarily less well used or less available for repurposing.

The effects of these associations are felt in urban practice. For instance, audiences are perhaps more inclined to receive greening projects as universal than they are other kinds of public goods. While Phoenix Lake prompted less planning, discussion, and controversy than a transparently social initiative might have, when New York mayor Michael Bloomberg proposed a West Side stadium as an economic development project in 2001 to fill the analogous "enormous, gaping hole" of the Hudson Yards—an infrastructural legacy of the industrial era—the public revolted, raising questions about the scale, utility, public funding, and public benefits of the project (Horowitz 2004). But Bloomberg's 2004 decision to support the privately initiated High Line—adapting another piece of obsolete industrial infrastructure, this time as a green economic development project for the Far West Side (Rainey 2014)—incited relatively minimal conflict and ultimately moved forward, subsidized by public dollars. The High Line was attractive because it capitalized on another powerful social imaginary, that of (high) culture, in promising galleries and public design as well as access to nature. But, as with Phoenix Lake, it was not until long after it was designed and built that project leaders began to fully reckon with its limited public uses and the changes it catalyzed in the neighborhood (Bliss 2017; Lindner and Rosa 2017; Reichl 2016). And, as with Phoenix Lake, project leaders now say that they should have anticipated these outcomes and included affordable housing as part of the design, and they are taking steps to make the space more accessible (Bliss 2017).

Urbanized nature also makes people willing to tolerate compromises made in and around green spaces that they might not with other kinds of projects, by suggesting that accessing nature is a universal—basic, human—experience impervious to social differences or inequalities built into the landscape. The description of Phoenix Lake as a "win-win" suggested that the lake was understood to be able to serve as a public good and a consumer good simultaneously; that the sale of adjacent houses would not compromise the experience of the lake. Compare this to a recent outrage provoked in New York City by so-called poor doors—separate entrances for subsidized, low- and moderate-income tenants of new luxury apartment buildings. While market-rate tenants have doormen, well-appointed entrances, and on-site spas, gyms, and private green space, subsidized tenants have a different experience of the same building. Because they do not pay for these amenities, they have separate entrances, no doormen, no gym or open space access, and no dishwashers or built-in light fixtures in their apartments. New Yorkers argued forcefully that no one should be subjected to this differential experience of the same space, that no child should be told that she cannot use the courtyard or must use a different entrance than her neighbor. But this is exactly the kind of income-stratified experience being pursued for public green spaces, as they are increasingly subsidized by proximate elite consumption. Why are these experiences understood to be damaging and discriminatory in the case of housing but acceptable for public parks? In part, the "good" of nature sustains the belief that individual experience is unaffected by such compromises in the case of green space.

## LESSONS FOR PRACTICE

In 2016, a short opinion piece in *The Guardian* comparing public responses to two proposed new green spaces in New York and London—New York City's Pier 55 and London's Garden Bridge—began: "There's something wonderfully uncontroversial about a park. A park can't be bad. We love trees. We love water. We love sunshine and flowers. Cities need open space, right? That built-in pleasure response means people are less likely to think of a park's cost, and to see only the leaves and grasses rather than the concrete and steel beneath them. They are also less likely to think of a park as precluding other, future uses." The author was interested in the very different public reception the two projects—"both offered as gifts to the city, both privately funded with a small public contribution, both backed by celebrities and mayors"—had received. New Yorkers largely supported Pier 55, while Londoners eventually saw Garden Bridge as a "land grab." The two proposed parks, both designed by the landscape architect Thomas Heatherwick, famous for his fantastic green renderings, were beautiful but controversial, their "questionable benefits and hidden costs" raising fundamental questions about democracy, access, and the price and distribution of public goods. The author concludes that New York's longer history of popular, privately funded public green spaces—from Central Park to the High Line—accounts for the difference in reception. Today, the Garden Bridge has been scrapped, but Pier 55 may still become a reality (Lange 2016).

The conceit of the *Guardian* piece and the conflict it explores are beautiful illustrations of the main arguments of this book. Regarding the power of nature, parks are "wonderfully uncontroversial," of incontestable benefit to cities. In terms of their motivation, they are aspirational public projects on the part of local elites who, much as Krupp and Osthaus did, want their cities to be rich in leisure and cultural amenities as well as economic power. Like the Ruhr's *Revierparks* and colonies in the 1970s, their virtue as public goods is largely communicated through their aesthetic greenness in planners' renderings and in the press surrounding the projects. Though both Pier 55 and Garden Bridge were—like Margarethenhöhe and IBA's Duisburg-Nord—offered as "gifts" to the cities, both were also more complicated. Critics argued that their revenue-generating private functions would chip away at open hours and public space; both parks would also still require significant public funding for construction and maintenance. And, as in the case of Phoenix Lake, both projects reflect a willingness on the part of receiving audiences to accept these outcomes as "unequal but beneficial" (Davidson 2014).

What are the implications of adopting this point of view for those who use, design, advocate for, and program green spaces—or simply those advocating for social justice and improved quality of life in cities? The first lesson is to *green better*, that is, with more reflexivity about the practice and its limitations. Knowing that greening projects cannot help but be partial accomplishments does not mean that we should simply accept their shortcomings; rather, it means that we should understand greening as the social project that it is and take what steps we can to democratize green spaces' planning, placement, conventions for use, and expectations for citizenship and urban practice, just as we commonly do in other areas of civic life. One legacy of the environmental justice movement is that scholars, activists, and practitioners have a much better-developed language for (and strategies for dealing with) the unequal distribution of environmental *bads* than *goods* (Gould and Lewis 2016; Park and Pellow 2011). We have a fairly good understanding of the disproportionate exposure of poor communities of color to toxic and polluting facilities, but a relatively impoverished vocabulary for talking about equal access to environmental amenities beyond their spatial distribution, especially regarding how environments understood to be natural and, therefore, universally accessible may be deeply unequal by virtue of classed, raced notions of nature and norms of public space use. This includes improving physical access but, equally importantly, improving social or cultural access by becoming more cognizant of parks as "environmental hegemonies"—articulations of state (or other forms of) power in the built environment (Sevilla-Buitrago 2017)—as activists, parents, and citizens are accustomed to doing with schools or courts or the media.

This is already becoming part of urban-environmental scholarship and professional practice. Scholars have begun to show how notions of race and class and ideals of citizenship and belonging get built in and policed in ostensibly public green spaces (e.g., Byrne 2012; and Newman 2015). In 2015, the centennial anniversary of the US national park system, a series of op-eds and articles asked "Why Are Our Parks So White?" and called on the National Park Service to carry out targeted outreach and diversity campaigns to overcome the trepidation, discrimination, and expense that racialized communities may associate with parks and wilderness areas and continue to experience in them (Finney 2014; Nelson 2015), while increasingly, city parks departments explicitly take immigrant park users and changing park use as part of their public mandate.

A second lesson is to *green less*, that is, to resist the urge to reach for green as a fix for all manner of social and ecological problems in the first place. It is hard to get outside this social imaginary, which governs thought and directs action. Even when there are obvious gaps between normative assertions that "everybody loves nature" and actual, varied experiences of green urban environments, rejecting the green-as-good narrative requires swimming upstream against the current of action that this social imaginary compels. As an example of the momentum of such beliefs, consider a study of trees in American cities, which found that "a normative discourse whereby trees are . . . represented as loved by everybody" produces intensive maintenance work on the part of city arborists "so as to maintain the people's idea about the trees being good" rather than face public challenges to this logic (Braverman 2008, 2, 5). Yet moments when the green-as-good logic breaks down—when people reject nature—can help bring competing values and interests into focus. I was once a teaching assistant for a class in which students were asked to design a better social housing complex for people living alone. Unsurprisingly, the students' two main recommendations were to put in a backyard garden and to turn the adjacent gas station into a public park. An audience member at the final public critique asked whether the gas station was not likely a far more well-used and much-needed amenity than another pocket park in the gas desert of Lower Manhattan would be. Where else, he asked, would taxi drivers find a bathroom? The students were flummoxed by the question—as was I, though by the fact that no critique of greening had surfaced until these final moments.

Antigreening arguments only sometimes actively question the underlying green-as-good assumption driving impassioned debates about access and distribution. For instance, one contemporary challenge to greening's social hegemony has come in the form of resistance to a phenomenon that has come to be known as *green gentrification*: the rising property values and displacement that often follow green upgrading in the context of hypercommodified housing markets and the widespread privatization of public resources. As with London's Garden Bridge, cities may be saying no to green gifts more often and poor and minority communities increasingly organizing against greening because they fear the gentrification that so often results (Checker 2011; Gould and Lewis 2016). Others have proposed making urban neighborhoods "just green enough" to be livable but not to displace (Curran and Hamilton 2017). In urban forestry, a discourse of "ecosystem disservices" is crystallizing whereby professionals highlight the downsides of urban trees and green spaces. Beyond taking stands against parks and trees, we might hope that scholars and community groups will begin to advocate for affordable housing *and* parks, and scale up from particular neighborhood battles to draw attention to the broader structural factors coproducing these dynamics. But, more to the point, these examples illustrate the need to be sensitive to the degree to which greening warps public politics. The hegemony of urbanized nature can cause us to forget to consider other kinds of solutions, such as public bathrooms. It can affect public decision-making processes, such as that favorite measure for improving access to public goods, community input. If asked, most communities will support increasing green space in most cases, but this will not tell planners much about other kinds of needs or how accessible that new green space is likely to be.

Most simply, this book draws attention to the deep roots and uncommon power of the complex of understandings surrounding nature in order to underscore how deeply entrenched and durable these imaginaries remain today. Making these sensibilities visible has a new urgency in the present. Common wisdom has it that tackling climate change will require a fundamental transformation of society and of society-nature relationships, and, at the moment, there is significant momentum around thinking about what these transformations should and could entail. Planning scholars and practitioners envision reconfigured spatial relationships, setting out new forms of dense, walkable urbanism or calling for just, carbon neutral cities. Urban farms and community gardens model new forms of sustainable agriculture and food access. Scholar-activists forge alliances between traditional environmental movements and social justice movements. Yet this book argues that underlying all this promising new thinking are a set of assumptions that are perhaps harder to think about and that are less often talked about: the degree to which a nineteenth-century imaginary remains entrenched in contemporary urban-environmental thinking and shapes orientations toward solutions. Take the bias toward green aesthetics in the sustainable urban landscapes presently being constructed. While buildings' color and plantings have a completely variable relationship to ecological outcomes, urban publics and sometimes professionals are prone to take such design choices as indicators of sustainability (Wachsmuth and Angelo 2018), overlooking overall negative environmental effects (Mössner and Miller 2015; Wachsmuth, Cohen, and Angelo 2016), while lower-carbon, more equitable solutions such as investment in affordable housing and public transportation remain less recognizable as forms of sustainable urbanism (Cohen 2017).

In the context of widespread agreement that addressing climate change will require myriad fundamental socioecological transformations, this analysis introduces another register in which such transformations must take place. It suggests that, in addition to remaking physical environments, forms of production and provisioning, and social movements and political alliances, transforming society-nature relationships will require unseating a powerful naturalized belief in the social value of nature. This will involve the creation of fundamentally new material and affective engagements with the environment and its signifiers—changes in human relationships to nature that are as significant and profound as were those that followed the transformations of industrial urbanization that first produced this imaginary. What would such a new imaginary look like? How would we get there? Might new environmental experiences, such as those of climate change, already be producing it? This book offers one way to start thinking outside urbanized nature by historicizing this imaginary and by situating it in relationship to large-scale social transformation. The hope is that bringing into focus the ideas that have driven greening in its many guises from the nineteenth century to the present, as well as the consequences of these practices, makes it possible to ask what we would want a new imaginary to be, what would characterize it, and how it might actually be produced.

ACKNOWLEDGMENTS

This project began in a previous life when, as an employee of New York City's Department of Parks and Recreation, I first became interested in the pervasive association of green with good and the zeal with which it allowed people to act. New York City's park volunteers, whose passion for picking up trash and planting street trees and daffodil bulbs—actions premised on the social value of green space, driven by the idea that caring for city parks and green streets improves urban neighborhoods—first attuned me to the morality of greening and the extent to which the green-as-good logic infuses contemporary urban life. Attempts to reckon with the occasional outlier—business- or homeowners who did not want a tree in front of their stores or homes and who reacted by committing "arborcide," killing the tree with bleach or soapy water—only reinforced how dominant this idea usually is. I am grateful to these volunteers for their tireless commitment to the city's public spaces and to my former colleagues at Partnerships for Parks, especially Whitney Files, Tamara Greenfield, Helen Ho, Emily Maxwell, and Neerja Vasishta, for puzzling through with me the many complexities of urban parks that eventually drove me to write this book. It was also at Parks that I first learned about the Ruhr's IBA Emscher Park project, which was, at the time, just gaining international attention.

When I left government for academe, it became obvious that, while park volunteers were not the only ones afflicted with a love of urban green space, only a university environment was tolerant of a decade-long journey to explain it. At New York University, I was lucky to find a fantastic group of interlocutors, role models, and sources of material and moral support. Craig Calhoun, in particular, made the years of fieldwork in Germany possible and has remained closely engaged with this project from inception to completion, sharpening it with his remarkable clarity of thought, insistence on intellectual seriousness and ambition, and unparalleled depth of knowledge. Neil Brenner shaped the foundations of my thinking on these topics and fostered the intellectual circle in which many of these ideas first germinated. He also showed incredible kindness and intellectual generosity in the first awkward transitional days from public-sector employee to academic and has done so many times since. I now cannot believe that Gianpaolo Baiocchi did not arrive at NYU until the end of my time there, so impossible is it to imagine this project without him. He may be the only person in the world who has read every word of this book, often twice. From the beginning, Colin Jerolmack shared my fascination with the allure of nature, and he has offered practical guidance, much-needed espressos, and many good ideas, only some of which I have been able to develop. It was Richard Sennett's work that initially sparked my interest in urbanism, many years before we met, and he maintained it with his impatience with disciplinary puzzle solving and orientation toward public intellectualism.

For their sagacity and camaraderie as graduate students and, now, as colleagues and collaborators, I am also especially grateful to David Wachsmuth and Daniel Aldana Cohen, along with Stuart Schrader, Aaron Jakes, and David Madden. The NYLON research network, Matthew Gandy, Manu Goswami, Harvey Molotch, Dierdre Royster, Iddo Tavory, Max Besbris, Max Holleran, Caitlin Petre, Anna Skarpelis, and Harel Shapira all provided excellent advice at various stages of the project, as did an indefatigable writing group composed of Gemma Mangione, Hannah Wohl, David Reinecke, and Stacy Williams.

The decision to carry out a historical project in Germany was, arguably, questionable. For this Sasha Disko-Schmidt is almost entirely responsible. She and Christopher Schmidt offered the best possible introduction to Berlin, as well as much-needed friendship and equally needed assistance demystifying many German systems, from Frakturschrift to the Ausländerbehörde. I am also grateful to Stefan Höhne, Boris Vormann, Christine Hentschel, Maren Freudenberg, Michael Kleinod, and Patricia Schulz, all of whom helped make Berlin a second home. In Dortmund, I thank Susanne Frank, Nina Schuster, and Anne Volkmann for their great hospitality and patience with my German, as well as Ingo Bader and family, Stefan Berger, and Thomas Urban for generously sharing their knowledge of the region. At various stages, Cordula Brinkmann, Lucas Stratmann, and Anne Kleinbauer provided indispensable and capable research assistance. I am also grateful for the hospitality of the Graduate School of Social Sciences at Humboldt University in Berlin and the

Department of Spatial Planning at TU Dortmund, each of which hosted me for a year, and for funding I received from the American Council of Learned Societies and the Horowitz Foundation for Social Policy.

I have been fortunate to complete much of the final manuscript at the University of California, Santa Cruz, surrounded by new forms of nature and urbanization that have caused me to rethink many of the project's premises. For their friendship and intellectual engagement, I am grateful to fellow Santa Cruzans Muriam Haleh Davis, Lindsey Dillon, Bill Domhoff, Deborah Gould, Lizzy Kate Gray, Miriam Greenberg, John Hall, Jamie Lyons, Thomas Serres, Matt Werner, and Michele Whizin. I was lucky to receive a Berggruen Fellowship to spend the 2017–18 academic year at New York University's Institute for Public Knowledge, where Eric Klinenberg, Harvey Molotch, and (again!) Gianpaolo Baiocchi all read chapters and, in some cases, the entire manuscript and continued to engage with these ideas. The final stages of the project also owe much to conversations with Claudio Benzecry, Andrew Deener, Kian Goh, Wendy Griswold, Terence McDonnell, Michael McQuarrie, Rachel Meier, Mariana Mogilevich, Garrett Dash Nelson, and Marie Warsh as well as to engaged audiences at UCSC's Cultural Studies and Environmental Studies colloquia and at Berkeley, Dartmouth, Georgetown, Northwestern, University College London, and the International Seminar on Urbanism and Urbanization at the University of Ghent. Some material in the book has been previously published in "The Greening Imaginary: Urbanized Nature in Germany's Ruhr Region," *Theory and Society* 48, no. 5 (2019): 645–69; "Green and Grey: New Ideologies of Nature in Urban Sustainability Policy," *Annals of the Association of American Geographers* 108, no. 4 (2018): 1038–56 (with David Wachsmuth); and "More Than Exception: Categories and the Problem of 'Nature' in the Ruhr," *Moving the Social* 50 (2013): 7–24.

At the University of Chicago Press, I am grateful to Doug Mitchell, who believed in this project at the beginning, and to Elizabeth Branch Dyson, Mollie McFee, and Christine Schwab, who saw it through at the end. The manuscript also greatly benefited from the comments of two anonymous reviewers.

A circle of friends and family also helped make this book possible. Richard Angelo, who models the writing life better than anyone I know. Linda Angelo, who, along with Larry Porter, reliably calls and sends cartoons when most needed. We all miss Merle Rosen, who could be counted on to tell us when Mercury was going retrograde. Dwight Billings, Karen Tice, Wallis Miller, and Frank Döring have been the best set of academic aunts and uncles an only child could ask for and have included me in their conversations for the past thirty years. Friendships with Catherine Cole Cimino, Samantha Freeman

Ebel, Jill Reinhardt, Chris Maxwell Rose, and Neerja Vasishta are now all, outrageously, decades old, and I am thrilled that none has been worn thin by my regular use of their guest rooms and couches. Finally, I am grateful to Ray Daniels, who has never known me without this project, and who has, over ten years and across three time zones, insisted on a life of humor, adventure, excellent cocktails, and sitting down to dinner.

# REFERENCES

Abend, Gabriel. 2014. *The Moral Background: An Inquiry into the History of Business Ethics*. Princeton, NJ: Princeton University Press.
Abrams, Lynn. 1992. *Workers' Culture in Imperial Germany*. London: Routledge.
Adams, Neil, and Philip Pinch. 2011. "What Is It about the IBA Emscher Park?" In *Internationale Bauausstellung Emscher Park: Impulse*, ed. Christa Reicher, Lars Niemann, and Angela Uttke, 184–97. Essen: Klartext.
Aglietta, Michel. 2000. *A Theory of Capitalist Regulation: The US Experience*. New York: Verso.
Albers, Gerd. 2006. "Urban Development, Maintenance and Conservation: Planning in Germany—Values in Transition." *Planning Perspectives* 21, no. 1:45–65.
Almond, Gabriel Abraham, and Sidney Verba. (1963) 2015. *The Civic Culture*. Reprint, Princeton, NJ: Princeton University Press.
Anderson, Benedict. 2006. *Imagined Communities*. New York: Verso.
Angelo, Hillary. 2017. "From the City as a Lens to Urbanization as a 'Way of Seeing': Country/City Binaries on an Urbanizing Planet." *Urban Studies* 54, no. 1:158–78.
Angelo, Hillary, and Kian Goh. 2020. "Out in Space: Difference and Abstraction in Planetary Urbanization." *International Journal of Urban and Regional Research*. https://doi.org/10.1111/1468-2427.12911.
Angelo, Hillary, and Colin Jerolmack. 2012. "Nature's Looking-Glass." *Contexts* 11, no. 1:24–29.
Angelo, Hillary, and David Wachsmuth. 2015. "Urbanizing Urban Political Ecology: A Critique of Methdological Cityism." *International Journal of Urban and Regional Research* 39, no. 1:16–27.
———. 2020. "Why Does Everyone Think Cities Can Save the Planet?" In "Why Does Everyone Think Cities Can Save the Planet," ed. Hillary Angelo and David Wachmuth, special issue of *Urban Studies*, vol. 57, no. 11:2201–21. https://doi.org/10.1177/0042098020919081.
Anguluri, Ramesh, and Priya Narayanan. 2017. "Role of Green Space in Urban Planning: Outlook towards Smart Cities." *Urban Forestry and Urban Greening* 25:58–65.

Arboleda, Martín. 2016. "Spaces of Extraction, Metropolitan Explosions: Planetary Urbanization and the Commodity Boom in Latin America." *International Journal of Urban and Regional Research* 40, no. 1:96–112.

Bahrdt, Hans-Paul. 1952. "Nachbarschaft oder Urbanität." *Bauwelt* 51/52:1467–77.

———. (1961) 2006. *Die moderne Großstadt*. Reprint, Wiesbaden: Verlag für Sozialwissenschaften.

———. 1968. *Humaner Städtebau: Überlegungen zur Wohnungspolitik und Stadtplanung für eine nahe Zukunft*. Vol. 4, *Zeitfragen*. Hamburg: Christian Wegner Verlag.

Baiocchi, Gianpaolo, Elizabeth Bennett, Alissa Cordner, Peter Taylor Klein, and Stephanie Savell. 2013. *The Civic Imagination*. New York: Paradigm.

Bargheer, Stefan. 2018. *Moral Entanglements: Conserving Birds in Britain and Germany*. Chicago: University of Chicago Press.

Barndt, Kerstin. 2010. "Memory Traces of an Abandoned Set of Futures: Industrial Ruins in the Post-Industrial Landscapes of Germany." In *Ruins of Modernity*, ed. Julia Hell and Andreas Schönle, 270–93. Durham, NC: Duke University Press.

Bartholomae, Florian, and Chang Woon Nam. 2014. "Are Large German Cities Really Shrinking? Demographic and Economic Development in Recent Years." In *Shrinking Cities: A Global Perspective*, ed. Harry W. Richardson and Chang Woon Nam, 86–104. New York: Routledge.

Baudrillard, Jean. (1998) 2016. *The Consumer Society: Myths and Structures*. Thousand Oaks, CA: Sage.

Beitz, Else. 1994. *"Das wird gewaltig ziehen und Früchte tragen!": Industriepädogogik in den Großbetrieben des 19. Jahrhunderts bis zum Ersten Weltkrieg dargestellt am Beispiel der Firma Fried. Krupp*. Essen: Klartext.

Bell, Michael Mayerfeld. 1995. *Childerley: Nature and Morality in a Country Village*. Chicago: University of Chicago Press.

———. 2018. *City of the Good: Nature, Religion, and the Ancient Search for What Is Right*. Princeton, NJ: Princeton University Press.

Bender, Thomas. 1982. *Toward an Urban Vision: Ideas and Institutions in Nineteenth Century America*. Baltimore: Johns Hopkins University Press.

Benedict, Andreas. 2000. *80 Jahre im Dienst des Ruhrgebiets: Siedlungsverband Ruhrkohlenbezirk (SVR) und Kommunalverband Ruhrgebiet (KVR) im historischen Überblick, 1920–2000*. Essen: Klartext.

Berger, Bennett. 1961. "The Myth of Suburbia." *Journal of Social Issues* 17, no. 1:38–49.

Berger, Stefan. 2012. *Germany*. London: Bloomsbury Academic.

Berger, Stefan, Jana Golombek, and Christian Wicke. 2018. "A Post-Industrial Mindscape? The Mainstreaming and Touristification of Industrial Heritage in the Ruhr." In *Industrial Heritage and Regional Identities*, ed. Christian Wicke, Stefan Berger, and Jana Golombek, 74–94. New York: Routledge.

Bergerhoff, Hans. 1928. *Untersuchungen über die Berg- und Rauchschädenfrage mit besonderer Berücksichtigung des Ruhrbezirks*. Godesberg and Bonn: Landwirtschaftliche Hochschule Bonn-Poppelsdorf.

Berman, Marshall. 1983. *All That Is Solid Melts into Air: The Experience of Modernity*. New York: Verso.

Blackbourne, David, and Geoff Eley. 1984. *The Peculiarities of German History: Bourgeois Society and Politics in Nineteenth-Century Germany.* Oxford: Oxford University Press.

Bliss, Laura. 2017. "The High Line's Next Balancing Act." February 7. Citylab. https://www.citylab.com/solutions/2017/02/the-high-lines-next-balancing-act-fair-and-affordable-development/515391.

Boltanski, Luc, and Laurent Thévenot. 2006. *On Justification: Economies of Worth.* Princeton, NJ: Princeton University Press.

Bolz, Cedric. 2010. "From 'Garden City Precursors' to 'Cemeteries for the Living': Contemporary Discourse on Krupp Housing and *Besucherpolitik* in Wilhelmine Germany." *Urban History* 37:90–116.

Borsdorf, Ulrich, and Sigrid Schneider. 2005. "A Mighty Business: Factory and Town in the Krupp Photographs." In *Pictures of Krupp: Photography and History in the Industrial Age*, ed. Klaus Tenefelde, 123–58. London: Philip Wilson.

Bösel, Monika. 1974. "Wie wohnen Arbeiter? Arbeiterwohnungen und Arbeiterviertel in der Bundesrepublik." *Der Bürger im Staat: Wohnen* 2:124–28.

Boström, Jörg, and Roland Günter. 1976. *Arbeiterinitiativen im Ruhrgebiet.* West Berlin: Verlag für das Studium der Arbeiterbewegung.

Bourdieu, Pierre. 1977. *Outline of a Theory of Practice.* New York: Cambridge University Press.

Brandi, D. 1912. "Die Margarethe-Krupp-Stiftung für Wohnungsfürsorge." In *Essens Entwicklung, 1812–1912, herausgegeben aus Anlaß der hundertjährigen Jubelfeier der Firma Krupp*, 43–46. Essen: Fredebeul & Koenen.

Braverman, Irus. 2008. "Everybody Loves Trees: Policing American Cities through Street Trees." *Duke Environmental Law and Policy Forum* 19:81–118.

Brenner, Neil. 2000. "Building 'Euro-Regions': Locational Politics and the Political Geography of Neoliberalism in Post-Unification Germany." *European Urban and Regional Studies* 7, no. 4:319–45.

———. 2004. *New State Spaces: Urban Governance and the Rescaling of Statehood.* New York: Oxford University Press.

———. 2013. "Theses on Urbanization." *Public Culture* 25, no. 1:85–114.

———, ed. 2014. *Implosions/Explosions: Towards a Study of Planetary Urbanization.* Berlin: Jovis.

Brenner, Neil, and Nikos Katsikis. 2014. "Is the Mediterranean Urban?" In *Implosions/Explosions: Towards a Study of Planetary Urbanization*, ed. Neil Brenner, 428–59. Berlin: Jovis.

Brenner, Neil, Jamie Peck, and Nik Theodore. 2010. "Variegated Neoliberalization: Geographies, Modalities, Pathways." *Global Networks* 10, no. 2:182–222.

Brenner, Neil, and Christian Schmid. 2015. "Towards a New Epistemology of the Urban?" *City* 19, nos. 2–3:151–82.

Brenner, Neil, and David Wachsmuth. 2012. "Territorial Competitiveness: Lineages, Practices, Ideologies." In *Planning Ideas That Matter*, ed. Bishwapriya Sanyal, Lawrence J. Vale, and Christina D. Rosan, 179–206. Cambridge, MA: MIT Press.

Brepohl, Wilhelm. 1957. *Industrievolk im Wandel von der agraren zur industriellen Daseinsform, dargestellt am Ruhrgebiet.* Tübingen: Mohr.

Brown, Timothy Scott. 2013. *West Germany and the Global Sixties.* Cambridge: Cambridge University Press.

Brubaker, Rogers. 1992. *Citizenship and Nationhood in France and Germany*. Cambridge, MA: Harvard University Press.

Brubaker, Rogers, and Frederick Cooper. 2000. "Beyond 'Identity.'" *Theory and Society* 29, no. 1:1–47.

Brüggemeier, Franz-Josef. 1983. *Leben vor Ort*. Munich: C. H. Beck.

———. 1994. "Nature Fit for Industry: The Environmental History of the Ruhr Basin, 1940–1990." *Environmental History Review* 18, no. 1:35–54.

Brüggemeier, Franz-Josef, Mark Cioc, and Thomas Zeller, eds. 2005. *How Green Were the Nazis? Nature, Environment, and Nation in the Third Reich*. Athens: Ohio University Press.

Brüggemeier, Franz-Josef, and Thomas Rommelspacher. 1992. *Blauer Himmel über der Ruhr: Geschichte der Umwelt im Ruhrgebiet, 1840–1990*. Essen: Klartext.

Buhrow, Walter, and Rudolf Holtappel. 1975. *Oberhausen: Im Mittelpunkt der Mensch*. Duisburg: Mercator-Verlag Gert Wohlfarth.

Burgess, Ernest W. 1925. "The Growth of the City: An Introduction to a Research Project." In *The City*, ed. Robert E. Park, Ernest W. Burgess, and Roderick D. McKenzie, 47–62. Chicago: University of Chicago Press.

Byrne, Jason. 2012. "When Green Is White: The Cultural Politics of Race, Nature and Social Exclusion in a Los Angeles Urban National Park." *Geoforum* 43, no. 3:595–611.

Calhoun, Craig. 1992. "Introduction: Habermas and the Public Sphere." In *Habermas and the Public Sphere*, ed. Craig Calhoun, 1–50. Cambridge, MA: MIT Press.

———. 2002. "Imagining Solidarity: Cosmopolitanism, Constitutional Patriotism, and the Public Sphere." *Public Culture* 14, no. 1:147–71.

———. 2008. "Cosmopolitanism in the Modern Social Imaginary." *Daedalus* 137, no. 3:105–14.

Calhoun, Craig, Dilip Gaonkar, Benjamin Lee, Charles Taylor, and Michael Warner. 2015. "Modern Social Imaginaries: A Conversation." *Social Imaginaries* 1, no. 1:189–224.

Čapek, Stella M. 2010. "Foregrounding Nature: An Invitation to Think about Shifting Nature-City Boundaries." *City and Community* 9, no. 2:208–24.

Carroll[-Burke], Patrick. 1996. "Science, Power, Bodies: The Mobilization of Nature as State Formation." *Journal of Historical Sociology* 9, no. 2:139–67.

———. 2002. "Material Designs: Engineering Cultures and Engineering States—Ireland, 1650–1900." *Theory and Society* 31, no. 1:75–114.

Castoriadis, Cornelius. 1997. *The Imaginary Institution of Society*. Cambridge, MA: MIT Press.

Checker, Melissa. 2011. "Wiped Out by the 'Greenwave': Environmental Gentrification and the Paradoxical Politics of Urban Sustainability." *City and Society* 23, no. 2:210–29.

Church, Roy, and Quentin Outram. 2002. *Strikes and Solidarity: Coalfield Conflict in Britain, 1889–1966*. Cambridge: Cambridge University Press.

Cioc, Mark. 2002. *The Rhine: An Eco-Biography, 1815–2000*. Seattle: University of Washington Press.

Cohen, Daniel Aldana. 2017. "The Other Low-Carbon Protagonists: Poor People's Movements and Climate Politics in Sao Paulo." In *The City Is the Factory*, ed. Miriam Greenberg and Penny Lewis, 140–57. Ithaca, NY: Cornell University Press.

Cohen, Lizabeth. 2003. *A Consumers' Republic: The Politics of Mass Consumption in Postwar America*. New York: Vintage.

Collins, George R., and Christiane Crasemann Collins. 1965. *Camillo Sitte and the Birth of Modern City Planning*. London: Phaidon.
Crawford, Margaret. 1995. *Building the Workingman's Paradise*. London: Verso.
Crew, David F. 1979. *Town in the Ruhr*. New York: Columbia University Press.
Cronon, William. 1992. *Nature's Metropolis: Chicago and the Great West*. New York: Norton.
———. 1995. "The Trouble with Wilderness; or, Getting Back to the Wrong Nature." In *Uncommon Ground: Rethinking the Human Place in Nature*, ed. William Cronon, 69–90. New York: Norton.
Curran, Winifred, and Trina Hamilton. 2017. *Just Green Enough: Urban Development and Environmental Gentrification*. New York: Routledge.
Danielzyk, Rainer. 1992. "Gibt es im Ruhrgebiet eine 'Postfordistische Regionalpolitik'?" *Geographische Zeitschrift* 80, no. 2:84–105.
Danielzyk, Rainer, and Gerald Wood. 2004. "Innovative Strategies of Political Regionalization: The Case of North Rhine-Westphalia." *European Planning Studies* 12, no. 2:191–207.
Daston, Lorraine, and Fernando Vidal, eds. 2004. *The Moral Authority of Nature*. Chicago: University of Chicago Press.
Davidson, Justin. 2014. "Barry Diller's Plan for a Floating Park Has De Blasio Channeling His Inner Bloomberg." November 17. Vulture. http://www.vulture.com/2014/11/rich-irony-of-barry-dillers-floating-park.html.
Dege, Wilhelm, and Wilfried Dege. 1983. *Das Ruhrgebiet*. Berlin: Gebrüder Borntraeger.
Dettmar, Jörg. 2005. "Forests for Shrinking Cities? The Project 'Industrial Forests of the Ruhr.'" In *Wild Urban Woodlands*, ed. Ingo Kowarik and Stefan Körner, 263–76. Berlin: Springer.
Diefendorf, Jeffry. 1990. *Rebuilding Europe's Bombed Cities*. New York: Palgrave Macmillan.
———. 1993. *In the Wake of War: The Reconstruction of German Cities after World War II*. New York: Oxford University Press.
———. 1999. "The West German Debate on Urban Planning." Paper presented at the conference "The American Impact on Western Europe: Americanization and Westernization in Transatlantic Perspective," German Historical Institute, Washington, DC, March 25–27. http://webdoc.sub.gwdg.de/ebook/p/2005/ghi_12/www.ghi-dc.org/conpotweb/western papers/diefendorf.pdf.
Dorsch, Petra. 1974. "Warum machen Trabantenstädte krank?" *Der Bürger im Staat: Wohnen* 2:120–23.
"Dortmund 'PHOENIX Lake.'" 2013. Werkstattstadt. https://www.nationale-stadtentwick lungspolitik.de/NSP/SharedDocs/Projekte/WSProjekte_ENG/Dortmund_PHOENIX _See.html.
Douglas, Mary. (1966) 2003. *Purity and Danger: An Analysis of Concepts of Pollution and Taboo*. Reprint, New York: Routledge.
Drucker, Peter. 2012. *Management*. New York: Routledge.
Dunlap, Riley E., and William R. Catton. 1994. "Struggling with Human Exemptionalism: The Rise, Decline and Revitalization of Environmental Sociology." *American Sociologist* 25, no. 1:5–30.
Eckart, Karl, Hartmut Kowalke, and John Mazeland. 2003. *Social, Economic, and Cultural Aspects in the Dynamic Changing Process of Old Industrial Regions*. Münster: Lit Verlag.

Eley, Geoff, 1989. "Labor History, Social History, 'Alltagsgeschichte': Experience, Culture, and the Politics of the Everyday—a New Direction for German Social History?" *Journal of Modern History* 61, no. 2:297-343.

———. 1992. "Nations, Politics, and Political Cultures: Placing Habermas in the Nineteenth Century." In *Habermas and the Public Sphere*, ed. Craig Calhoun, 289-339. Cambridge, MA: MIT Press.

———. 2002. "Politics, Culture, and the Public Sphere." *Positions: East Asia Cultures Critique* 10, no. 1:219-36.

Eliasoph, Nina. 1998. *Avoiding Politics: How Americans Produce Apathy in Everyday Life*. Cambridge: Cambridge University Press.

Elliott, Anthony. 2004. *Social Theory since Freud: Traversing Social Imaginaries*. New York: Routledge.

———. 2012. "New Individualist Configurations and the Social Imaginary." *European Journal of Social Theory* 15, no. 3:349-65.

Emirbayer, Mustafa, and Anne Mische. 1998. "What Is Agency?" *American Journal of Sociology* 103, no. 4:962-1023.

Engelke, Peter Owen. 2011. "Green City Origins: Democratic Resistance to the Auto-Oriented City in West Germany, 1960-1990." PhD diss., Georgetown University.

Engels, Frederick. (1872) 1935. *The Housing Question*. Edited by Clemens Palme Dutt. New York: International.

Ernstson, Henrik, and Erik Swyngedouw, eds. 2018. *Urban Political Ecology in the Anthropo-Obscene: Interruptions and Possibilities*. London: Routledge.

Faecke, Peter, Rolf Stefaniak, and Gerd Haag. 1977. *Gemeinsam gegen Abriß: Ein Lesebuch aus Arbeitersiedlungen und ihren Initiativen*. Wuppertal: Peter Hammer.

Farrell, Justin. 2015. *The Battle for Yellowstone: Morality and the Sacred Roots of Environmental Conflict*. Princeton, NJ: Princeton University Press.

Fine, Gary Alan. 2009. *Morel Tales*. Cambridge, MA: Harvard University Press.

Finney, Carolyn. 2014. *Black Faces, White Spaces: Reimagining the Relationship of African Americans to the Great Outdoors*. Chapel Hill: University of North Carolina Press.

Fischer-Eckert, L. 1913. *Die wirtschaftliche und soziale Lage der Frauen in dem modernen Industrieort Hamborn im Rheinland*. Hagen: Carl Stracke.

FitzSimmons, Margaret. 1989. "The Matter of Nature." *Antipode* 21, no. 2:106-20.

Flagge, Ingeborg, ed. 1999. *Geschichte des Wohnens: Von 1945 bis heute—Aufbau—Neubau—Umbau*. Stuttgart: Deutsche Verlags-Anstalt.

Fourcade, Marion. 2011. "Cents and Sensibility: Economic Valuation and the Nature of 'Nature.'" *American Journal of Sociology* 116, no. 6:1721-77.

Frank, Susanne. 2010. "Rückkehr der Natur: Die Neuerfindung von Natur und Landschaft in der Emscherzone." EMSCHERplayer. http://www.emscherplayer.de/main.yum?mainActi on=magazin&id=49786.

Franklin, Deborah. 2012. "Nature That Nurtures." *Scientific American* 306, no. 3:24-25. Available online (as "How Hospital Gardens Help Patients Heal") at https://www.scientific american.com/article/nature-that-nurtures.

Fraser, Nancy. 1990. "Rethinking the Public Sphere: A Contribution to the Critique of Actually Existing Democracy." *Social Text* 25/26:56–80.
Friedan, Betty. 1963. *The Feminine Mystique*. New York: Norton.
Friedrichs, Jürgen. 1987. "Urban Renewal Policies and Back-to-the-City Migration: The Case of West Germany." *Journal of the American Planning Association* 53, no. 1:70–79.
Fritsch, Theodor. 1896. *Die Stadt der Zukunft*. Leipzig: Hammer.
Frye, Margaret. 2012. "Bright Futures in Malawi's New Dawn: Educational Aspirations as Assertions of Identity." *American Journal of Sociology* 117, no. 6:1565–1624.
Fuchs, Carl Johannes. 1908. "Die Gartenstadt (Auszug)." In *Im Grünen wohnen—im Blauen planen (1990)*, ed. Franziska Bollery, Gerhard Fehl, and Kristiana Hartmann, 105–10. Hamburg: Hans Christians Verlag.
Gaida, Wolfgang, and Helmut Grothe. 1997. *Vom Kaisergarten zum Revierpark*. Bottrop and Essen: Kommunalverband Ruhrgebiet/Verlag Peter Pomp.
Galvis, Juan. 2014. "Remaking Equality: Community Governance and the Politics of Exclusion in Bogota's Public Spaces." *International Journal of Urban and Regional Research* 38, no. 4:1458–75.
Gandy, Matthew. 2002. *Concrete and Clay: Reworking Nature in New York City*. Cambridge, MA: MIT Press.
———. 2013. "Marginalia: Aesthetics, Ecology, and Urban Wastelands." *Annals of the Association of American Geographers* 103:1301–16.
Ganser, Karl. 2010. "Art Is Wilderness's Closest Neighbor." In *Unter freiem Himmel: Emscher Landschaftspark*, ed. Regionalverband Ruhr, 9–13. Basel: Birkhäuser.
Gaonkar, Dilip Parameshwar. 2002. "Toward New Imaginaries: An Introduction." *Public Culture* 14, no. 1:1–19.
Gausmann, P., J. Weiss, P. Keil, and G. H. Loos. 2007. "Wildnis kehrt zurück in den Ballungsraum." *Praxis der Naturwissenschaften—Biologie in der Schule* 56, no. 2:27–32.
Geertz, Clifford. 1972. "Deep Play: Notes on the Balinese Cockfight." *Daedalus* 134, no. 4:56–86.
Gesellschaft für Heimkultur and Hermann Hecker. 1917. *Der Krupp'sche Kleinwohnungsbau*. Wiesbaden: Heimkultur-Verlagsgesellschaft mbH.
Gieryn, Thomas F. 2002. "What Buildings Do." *Theory and Society* 31:35–74.
Girardet, Herbert. 2014. *Creating Regenerative Cities*. New York: Routledge.
Glacken, Clarence J. 1967. *Traces on the Rhodian Shore: Nature and Culture in Western Thought from Ancient Times to the End of the Eighteenth Century*. Berkeley: University of California Press.
Glaeser, Andreas. 2011. *Political Epistemics*. Chicago: University of Chicago Press.
Goch, Stefan. 2002a. "Betterment without Airs: Social, Cultural, and Political Consequences of De-Industrialization in the Ruhr." *International Review of Social History* 47, no. 10:87–111.
———. 2002b. *Eine Region im Kampf mit dem Strukturwandel*. Essen: Klartext.
Godau, Sigrid, and Claudia Heinrich. 2010. "Zollverein Park." In *Unter freiem Himmel: Emscher Landschaftspark*, ed. Sabine Auer, Anna Margarethe Lavier, and Regionalverband Ruhr, 84–93. Basel: Birkhäuser.

Göderitz, J., R. Rainer, and H. Hoffman. 1957. *Die gegliederte und aufgelockerte Stadt*. Tübingen: Ernst Wasmuth.
Goffman, Erving. 2008. *Behavior in Public Places*. New York: Simon & Schuster.
Goswami, Manu. 2002. "Rethinking the Modular Nation Form: Toward a Sociohistorical Conception of Nationalism." *Comparative Studies in Society and History* 44, no. 4:770–99.
Gould, Kenneth, and Tammy Lewis. 2016. *Green Gentrification: Urban Sustainability and the Struggle for Environmental Justice*. New York: Routledge.
Govindarajulu, Dhanapal. 2014. "Urban Green Space Planning for Climate Adaptation in Indian Cities." *Urban Climate* 10:35–41.
Gramsci, Antonio. 1996. *Prison Notebooks*. Translated by Joseph A. Buttigieg. New York: Columbia University Press.
Grazian, David. 2015. *American Zoo: A Sociological Safari*. Princeton, NJ: Princeton University Press.
Green, Nicholas. 1990. *The Spectacle of Nature*. Manchester: Manchester University Press.
Greenberg, Miriam. 2015. "'The Sustainability Edge': Competition, Crisis, and the Rise of Green Urban Branding." In *Sustainability in the Global City*, ed. Cindy Isenhour, Gary McDonogh, and Melissa Checker, 105–30. New York: Cambridge University Press.
"Greenest Block in Brooklyn." n.d. Brooklyn Botanic Garden. https://www.bbg.org/community/greenestblock.
"The Greenest Block in Brooklyn (1902)." 2017. The Brownstone Detectives, July 10. http://www.brownstonedetectives.com/miss-zella-milhaus-greenest-block-in-brooklyn-1902.
Griswold, Wendy, Gemma Mangione, and Terence E. McDonnell. 2013. "Objects, Words, and Bodies in Space: Bringing Materiality into Cultural Analysis." *Qualitative Sociology* 36, no. 4:343–64.
Grothe, Helmut. 2003. "Die Revierparks im Ruhrgebiet—eine Erfolgsgeshichte." *Industriedenkmalpflege und Geschichtskultur* 1:38–41.
Grove, Richard H. 1996. *Green Imperialism: Colonial Expansion, Tropical Island Edens and the Origins of Environmentalism, 1600–1860*. Oxford: Cambridge University Press.
Gualini, Enrico. 2004. "Regionalization as 'Experimental Regionalism': The Rescaling of Territorial Policy-Making in Germany." *International Journal of Urban and Regional Research* 28, no. 2:329–53.
Günel, Gökçe. 2019. "Exporting the Spaceship: The Connected Isolation of Masdar City." In *The New Arab Urban: Gulf Cities of Wealth, Ambition, and Distress*, ed. Harvey Molotch and Davide Ponzini, 194–212. New York: New York University Press.
Günter, Janne. 1980. *Leben in Eisenheim*. Weinheim: Beltz.
Günter, Janne, and Roland Günter. 1999. *"Sprechende Straßen" in Eisenheim*. Essen: Klartext.
Günter, Roland. 1970. "Krupp und Essen." In *Das Kunstwerk zwischen Wissenschaft und Weltanschauung*, ed. Martin Warnke, 128–74. Gütersloh: Bertelsmann Kunstverlag.
———. 1977. "Eisenheim—das ist eine Art miteinander zu leben." In *Nachbarschaft im Neubaublock*, ed. Reimer Gronemeyer and Hans-Eckehard Bahr, 294–327. Weinheim and Basel: Beltz.
Günter, Roland, Janne Günter, and Leonard Henny. 1982. "Visual Essay: Save Eisenheim." *European Newsletter on Visual Sociology* 4:24–30.

Günter, Roland, and Michael Weisser. 1975. "The Workingmen's Colony at Eisenheim Near Oberhausen, West Germany." In *Transactions of the First International Congress on the Conservation of Industrial Monuments, Ironbridge, May 29–June 5, 1973*, 92–97. Ironbridge: Ironbridge Gorge Museum Trust.

Günther, Adolf, and René Prévôt. 1905. *Die Wohlfahrtseinrichtungen der Arbeitgeber in Deutschland und Frankreich.* Leipzig: Duncker & Humblot.

Gust, Claudia. 2008. "Abenteuerspielplatz am Phoenix-See." Playground+Landscape. http://www.playground-landscape.com/de/article/view/1084.html.

Haase, Dagmar. 2008. "Urban Ecology of Shrinking Cities: An Unrecognized Opportunity?" *Nature and Culture* 3, no. 1:1–8.

Habermas, Jürgen. 1962. *Strukturwandel der Öffentlichkeit.* Darmstadt and Neuwied: Hermann Luchterhand Verlag.

———. 1988. "Historical Consciousness and Post-Traditional Identity: Remarks on the Federal Republic's Orientation to the West." *Acta Sociologica* 31, no. 1:3–13.

———. 1991a. *The Structural Transformation of the Public Sphere.* Translated by Thomas Burger with the assistance of Frederick Lawrence. Cambridge, MA: MIT Press.

———. 1991b. "Yet Again: German Identity: A Unified Nation of Angry DM-Burghers?" *New German Critique* 52:84–101.

Hall, Peter. 1966. *The World Cities.* London: World University Library.

———. (1988) 1994. *Cities of Tomorrow: An Intellectual History of Urban Planning and Design in the Twentieth Century.* Oxford: Blackwell.

———. 2002. *Cities of Tomorrow.* Malden: Wiley-Blackwell.

Hancock, M. Donald. 1978. "Productivity, Welfare, and Participation in Sweden and West Germany: A Comparison of Social Democratic Reform Prospects." *Comparative Politics* 11, no. 1:4–23.

Hansen, Miriam. 1993. "Unstable Mixtures, Dilated Spheres: Negt and Kluge's *The Public Sphere and Experience*, Twenty Years Later." *Public Culture* 5, no. 2:179–212.

Harris, Teresa. 2012. "The German Garden City Movement: Architectur, Politics and Urban Transformation, 1902–1931." PhD diss., Columbia University.

Harvey, David. 1993. "The Nature of Environment: Dialectics of Social and Environmental Change." *Socialist Register* 29:1–51.

———. 2006. *Paris, Capital of Modernity.* New York: Routledge.

Heinrichsbauer, A. 1936. *Industrielle Siedlung im Ruhrgebiet.* Essen: Glückauf.

Helfrich, Andreas. 2000. *Die Margarethenhöhe Essen.* Weimar: VDG.

Hemmings, Sarah, and Martin Kagel. 2010. "Memory Gardens: Aesthetic Education and Political Emancipation in the 'Landschaftspark Duisburg-Nord.'" *German Studies Review* 33, no. 2:243–62.

Herbert, Ulrich. 1983. "Vom Kruppianer zum Arbeitnehmer." In *"Hinterher merkt man, dass es richtig war, dass es schiefgegangen ist": Nachkriegserfahrungen im Ruhrgebiet*, ed. Lutz Niethammer, 233–76. Bonn: J. H. W. Dietz.

Herlyn, Ulfert. 1970. *Wohnen im Hochhaus: Eine empirische-soziologische Untersuchung in ausgewählten Hochhäusern der Städte München, Stuttgart, Hamburg, und Wolfsburg.* Stuttgart: Karl Krämer.

———. 2006. "Zur Neuauflage des Buches *Die moderne Großstadt*." In *Die moderne Großstadt* (1961), by Hans-Paul Bahrdt, 7–26. Reprint, Wiesbaden: Verlag für Sozialwissenschaften.

Hesse-Frielinghaus, Herta, ed. 1971. *Karl Ernst Osthaus: Leben und Werk*. Recklinghausen: Aurel Bongers.

Heynen, Nikolas, Maria Kaika, and Erik Swyngedouw, eds. 2006. *In the Nature of Cities: Urban Political Ecology and the Politics of Urban Metabolism*. New York: Routledge.

Hezel, Dieter. 1974. "Kommunale Freizeitpolitik." *Der Bürger im Staat: Wohnen* 2:136–40.

Hickey, S. H. F. 1985. *Workers in Imperial Germany: The Miners of the Ruhr*. Oxford: Oxford University Press.

Historischen Vereins für Dortmund und die Grafschaft Mark e. V. in Verbindung mit dem Stadtarchiv Dortmund, ed. 2012. *"Treffpunkt Dortmund": Modernes Stadtmarketing einer Ruhrgebietsstadt im Strukturwandel, 1965–1990*. Essen: Klartext.

Hitlin, Steven, and Stephen Vaisey. 2013. "The New Sociology of Morality." *Annual Review of Sociology* 39:51–68.

Hobsbawm, Eric J. 1969. *Industry and Empire: From 1750 to the Present Day*. Pelican Economic History of Britain, vol. 3. Baltimore: Penguin.

Hoffmann, Hilmar. 1982. *Das Taubenbuch*. Frankfurt a.M.: W. Kruger.

Holm, Andrej, and Armin Kuhn. 2011. "Squatting and Urban Renewal: The Interaction of Squatter Movements and Strategies of Urban Restructuring in Berlin." *International Journal of Urban and Regional Research* 35, no. 3:644–58.

Honhart, Michael. 1990. "Company Housing as Urban Planning in Germany, 1870–1940." *Central European History* 23, no. 1:3–21.

Horowitz, Craig. 2004. "Stadium of Dreams." *New York Magazine*, June 11. http://nymag.com/nymetro/realestate/urbandev/features/9307.

Horst, Christiane. 1994. "Entwicklungskonzept für den Stadtgarten Essen." PhD diss., Universität-Gesamthochschule Essen.

Hospers, Gert-Jan. 2004. "Restructuring Europe's Rustbelt: The Case of the German Ruhrgebiet." *Intereconomics* 39, no. 3 (May/June): 147–56.

Hospers, Gert-Jan, and Burkhard Wetterau. 2018. *Metrople Ruhr—Small Atlas*. 3rd rev. ed. Essen: Regionalverband Ruhr.

Howard, Ebenezer. (1898) 2010. *To-Morrow: A Peaceful Path to Real Reform*. New York: Cambridge University Press.

———. 1902. *Garden Cities of To-Morrow*. London: S. Sonnenschein.

Huck, Gerhard, ed. 1980. *Sozialgeschichte der Freiheit*. Wuppertal: Peter Hammer.

Hudson, Ray, and Dan Swanton. 2011. "Global Shifts in Contemporary Times: The Changing Trajectories of Steel Towns in China, Germany, and the United Kingdom." *European Urban and Regional Studies* 19, no. 1:6–19.

Hundt, Robert. 1902. *Bergarbeiter-Wohnungen im Ruhrrevier*. Berlin: Julius Springer.

Huning, Sandra, and Susanne Frank. 2011. "Urban Waterscapes as Products, Media, and Symbols of Change—the Reinvention of the Ruhr." Paper presented at the EFLA regional congress, November 2–4, Tallinn. http://129.217.131.68:8080/bitstream/2003/35974/1/Huning_Frank_Urban%20Waterscapes%20as%20Products%2C%20Media%20and%20Symbols%20of%20Change_Tallinn%20paper.pdf.

Ives, Colta. 2018. *Public Parks, Private Gardens: Paris to Provence*. New York: Metropolitan Museum of Art.
Jackson, James H., Jr. 1997. *Migration and Urbanization in the Ruhr Valley, 1821–1914*. Boston: Humanities.
Jacobs, Jane. 1961. *The Death and Life of Great American Cities*. New York: Vintage.
Jameson, Fredric. 1988. "On Negt and Kluge." *October* 46:151–77.
Jantke, Carl. 1953. *Bergmann und Zeche*. Tübingen: Mohr.
Jelin, Elizabeth. 1974. "The Concept of Working-Class Embourgeoisement." *Studies in Comparative International Development* 9, no. 1:1–19.
Jencks, Charles. 1977. *The Language of Post-Modern Architecture*. New York: Rizzoli.
Jerolmack, Colin. 2007. "Animal Practices, Ethnicity, and Community: The Turkish Pigeon Handlers of Berlin." *American Sociological Review* 72, no. 6:874–94.
———. 2013. *The Global Pigeon*. Chicago: University of Chicago Press.
Jessop, Bob. 1994. "The Transition to Post-Fordism and the Schumpeterian Workfare State." In *Towards a Post-Fordist Welfare State?*, ed. Roger Burrows and Brian Loder, 13–37. London: Psychology Press.
———. 2002. *The Future of the Capitalist State*. Cambridge: Polity.
———. 2010. "Cultural Political Economy and Critical Policy Studies." *Critical Policy Studies* 3, nos. 3–4:336–56.
Jessop, Bob, and Ngai-Ling Sum. 2006. *Beyond the Regulation Approach: Putting Capitalist Economies in Their Place*. Northampton: Edward Elgar.
Jonas, Michael. 2008. "About an Urban Development Process: The Case of the Dortmund-Project." Institut für Höhere Studien Sociological Series Working Paper no. 87. Vienna: Institute for Advanced Studies.
———. 2013. "The Dortmund Case—on the Enactment of an Urban Economic Imaginary." *International Journal of Urban and Regional Research* 38, no. 6:2123–40.
Joyce, Patrick. 2003. *The Rule of Freedom: Liberalism and the Modern City*. London: Verso.
Juckel, Lothar. 1992. *Stadtbildprägende Arbeitersiedlungen: Erhalten und Erneuerung denkmalwerter Arbeitersiedlungen im Rhein-Ruhr-Gebiet*. Dortmund: Institut für Landes- und Stadtentwicklungsforschung des Landes Nordrhein-Westfalen.
Kahn, Joseph, and Mark Landler. 2007. "China Grabs West's Smoke-Spewing Factories." *New York Times*, December 21.
Kaika, Maria. 2005. *City of Flows: Modernity, Nature, and the City*. New York: Routledge.
Kallen, Peter W. 1984. "Idylle oder Illusion? Die Margarethenhöhe in Essen von Georg Metzendorf." In *Die Margarethenhöhe: Das Schöne und die Ware*, ed. Tünn Konerding and Zdenek Felix, 48–96. Essen: Museum Folkwang.
Kaschuba, Wolfgang. 1995. "Deutsche Bürgerlichkeit nach 1800: Kultur als symbolische Praxis." In *Bürgertum im 19. Jahrhundert: Wirtschaftsbürger und Bildungsbürger*, 2, ed. Jürgen Kocka, 92–127. Göttingen: Vandenhoeck & Ruprecht.
Kastorff-Viehmann, Renate. 1998. "Die Stadt und das Grün, 1860 bis 1960." In *Die grüne Stadt*, ed. Renate Kastorff-Viehmann, 49–142. Essen: Klartext.
Kastorff-Viehmann, Renate, and Yasemin Utku. 2012. "Das Erbe Robert Schmidts." *Raumplanung* 161, no. 2:47–48.

Keil, Andreas, and Burkhard Wetterau. 2013. *Metropolis Ruhr: A Regional Study of the New Ruhr*. Translated by Hans-Werner Wehling. Essen: Regionalverband Ruhr.

Keil, Roger. 2017. *Suburban Planet: Making the World Urban from the Outside In*. Hoboken, NJ: John Wiley & Sons.

Keil, Roger, and John Graham. 2005. "Reasserting Nature: Constructing Urban Environments After Fordism." In *Remaking Reality: Nature at the Millennium*, ed. Bruce Braun and Noel Castree, 100–125. London: Routledge.

Kellen, T. 1902. *Die Industriestadt Essen in Wort und Bild: Geschichte und Beschreibung der Stadt Essen*. Essen-Ruhr: Fredebeul & Koenen.

Kift, Dagmar. 2005. "Bergbau. Aufschwung mit Zuwanderern." In *Aufbau West: Neubeginn zwischen Vertreibung und Wirtschaftswunder*, ed. Dagmar Kift, 84–88. Essen: Klartext.

———. 2013. "Brass Bands and Beat Bands, Poets and Painters: A Cross-Cultural Case Study of Mining Culture and Regional Identity in the Ruhr Area, 1947–1966." *International Journal of Heritage Studies* 19, no. 5:495–510.

Kilper, Heiderose, and Gerald Wood. 1995. "Restructuring Policies: The Emscher Park International Building Exhibition." In *The Rise of the Rustbelt*, ed. Phil Cooke, 208–31. New York: St. Martin's.

Klapheck, Richard. 1930. *Siedlungswerk Krupp*. Berlin: Ernst Wasmuth.

Kleinert, Uwe. 1988. *Flüchtlinge und Wirtschaft in Nordrhein-Westfalen, 1945–1961: Arbeitsmarkt—Gewerbe—Staat*. Düsseldorf: Schwann.

Klemek, Christopher. 2007. "Placing Jane Jacobs within the Transatlantic Urban Conversation." *Journal of the American Planning Association* 73, no. 1:49–67.

———. 2011. *The Transatlantic Collapse of Urban Renewal*. Chicago: University of Chicago Press.

Klink, Heinz-Dieter, and Thomas Rommelspacher. 2009. "Nachdruck aus besonderem Anlass: Robert Schmidt und die Gründungsurkunde des SVR." In *Denkschrift betreffend Grundsätze zur Aufstellung eines General-Siedelungsplanes* (1912), by Robert Schmidt. Reprint, Essen: Klartext.

Klösters, Hans G. 1982. *Dichtung in Stein und Grün: Margarethenhöhe*. Essen: Margarethe Krupp-Stiftung für Wohnungsfürsorge.

Knabe, W., K. Mellinghoff, F. Meyer, and R. Schmidt-Lorenz. 1968. *Haldenbegrünung im Ruhrgebiet*. Essen: Siedlungsverband Ruhrkohlenbezirk.

Koch, Max Jürgen. 1954. *Die Bergarbeiterbewegung im Ruhrgebiet zur Zeit Wilhelms II*. Düsseldorf: Droste.

Koopmans, Ruud. 1995. *Democracy from Below: New Social Movements and the Political System in West Germany*. Boulder, CO: Westview.

Koshar, Rudy. 1998. *Germany's Transient Pasts: Preservation and National Memory in the Twentieth Century*. Chapel Hill: University of North Carolina Press.

Kramer, Lloyd. 1992. "Habermas, History, and Critical Theory." In *Habermas and the Public Sphere*, ed. Craig Calhoun, 236–58. Cambridge, MA: MIT Press.

Krause, Monika. 2006. "The Production of Counter-Publics and the Counter-Publics of Production: An Interview with Oskar Negt." *European Journal of Social Theory* 9, no. 1:119–28.

Krinsky, John, and Maud Simonet. 2017. *Who Cleans the Park?* Chicago: University of Chicago Press.

Krupar, Shiloh. 2013. *Hot Spotter's Report: Military Fables of Toxic Waste*. Minneapolis: University of Minnesota Press.

Krupp'sche Gussstahlfabrik. 1912a. *Krupp, 1812–1912: Zum 100-jährigen Bestehen der Firma Krupp und der Gussstahlfabrik zu Essen Herausgegeben auf den hundertsten Geburtstag Alfred Krupps*. Jena: Gustav Fischer.

———. 1912b. *Krupp: A Century's History of the Krupp Works. Translated from the Commemorative Volume Edited by the Krupp Works*. Essen.

Kruse, Wilfried, and Rainer Lichte, eds. 1991. *Krise und Aufbruch in Oberhausen: Zur Lage der Stadt und ihrer Bevölkerung am Ausgang der achtziger Jahre*. Oberhausen: ASSO.

Kunzig, Robert. 2017. "The World's Most Improbable Green City." April 4. National Geographic. https://www.nationalgeographic.com/environment/urban-expeditions/green-buildings/dubai-ecological-footprint-sustainable-urban-city.

Ladd, Brian. 1990. *Urban Planning and Civic Order in Germany, 1860–1914*. Cambridge, MA: Harvard University Press.

Landesregierung Nordrhein-Westfalen. 1968. *Entwicklungsprogramm Ruhr, 1968–1973*. Düsseldorf: Institut für Landes- und Stadtentwicklungsforschung des Landes Nordrhein-Westfalen.

———. 1970. *Nordrhein-Westfalen-Programm, 1975*. Düsseldorf: Institut für Landes- und Stadtentwicklungsforschung des Landes Nordrhein-Westfalen.

Lange, Alexandra. 2016. "Garden Bridge v Pier 55: Why Do New York and London Think So Differently?" *The Guardian*, May 26. https://www.theguardian.com/cities/2016/may/26/garden-bridge-pier-55-new-york-london.

Larsen, Clifford. 1996. "What Should Be the Leading Principles of Land Use Planning? A German Perspective." *Vanderbilt Journal of Transnational Law* 29, no. 5:967–1017.

Lears, T. J. Jackson. 1985. "The Concept of Cultural Hegemony: Problems and Possibilities." *American Historical Review* 90, no. 3:567–93.

Lees, Andrew. 1985. *Cities Perceived: Urban Society in European and American Thought, 1820–1940*. Manchester: Manchester University Press.

———. 2002. *Cities, Sin, and Social Reform in Imperial Germany*. Ann Arbor: University of Michigan Press.

Lefebvre, Henri. (1947) 1991. *Critique of Everyday Life*. Vol. 1, *Introduction*. Translated by John Moore. New York: Verso.

———. (1961) 2002. *Critique of Everyday Life*. Vol. 2, *Foundations for a Sociology of the Everyday*. Translated by John Moore. New York: Verso.

———. (1970) 2003. *The Urban Revolution*. Translated by R. Bononno. Minneapolis: University of Minnesota Press.

———. (1981) 2005. *Critique of Everyday Life*. Vol. 3, *From Modernity to Modernism*. Translated by Gregory Elliott. New York: Verso.

Lekan, Thomas M., and Thomas Zeller, eds. 2005. *Germany's Nature*. New Brunswick, NJ: Rutgers University Press.

Leschny-Kröger, Eva. 2010. "Zur Geschichte der Kleingärten im Ruhrgebiet." In *Zwischen Kappes und Zypressen: Gartenkunst an Emscher und Ruhr*, ed. Martina Oldengott and Christine Vogt, 59–62. Essen: Klartext.

Levenstein, Adolf. 1912. *Die Arbeiterfrage*. Munich: Ernst Reinhardt.

Levine, Alexandra S. 2017. "New York Today: The Greenest Block in Brooklyn." *New York Times*, July 31. https://www.nytimes.com/2017/07/31/nyregion/new-york-today-the-greenest-block-in-brooklyn.html?mtrref=undefined&gwh=84A4B55DDA0888CA5CFEB181B96F36AE&gwt=pay.

Lindner, Christoph, and Brian Rosa, eds. 2017. *Deconstructing the High Line*. New Brunswick, NJ: Rutgers University Press.

Lofland, Lyn. 1989. "The Morality of Urban Public Life: The Emergence and Continuation of a Debate." *Places* 6, no. 1:18–23.

———. 1998. *The Public Realm: Exploring the City's Quintessential Social Territory*. New Brunswick: Transaction.

Love, Shayla. 2014. "Here's the Greenest Block in Brooklyn." *Gothamist*, August 6. http://gothamist.com/2014/08/06/greenest_block_brooklyn.php#photo-4.

Lubow, Arthur. 2004. "The Anti-Olmsted." *New York Times Magazine*, May 16, 47–54.

Lüdtke, Alf, ed. 1995a. *The History of Everyday Life*. Translated by W. Templer. Princeton, NJ: Princeton University Press.

———. 1995b. "What Happened to the 'Fiery Red Glow'? Workers' Experiences and German Fascism." In *The History of Everyday Life*, ed. Alf Lüdtke, trans. W. Templer, 198–251. Princeton, NJ: Princeton University Press.

Lüdtke, Hartmut. 1972. *Freizeit in der Industriegesellschaft: Emanzipation oder Anpassung?* Opladen: Leske.

Lynch, Kevin. 1960. *The Image of the City*. Cambridge, MA: MIT Press.

Madden, David. 2010. "Revisiting the End of Public Space: Assembling the Public in an Urban Park." *City and Community* 9, no. 2:187–207.

Manchester, William. 1970. *The Arms of Krupp*. Boston: Bantam.

Mangione, Gemma. 2016. "Making Sense of Things: Constructing Aesthetic Experience in Museum Gardens and Galleries." *Museum and Society* 14, no. 1:33–51.

Margarethe Krupp-Stiftung für Wohnungsfürsorge. 1915. "Mietvertrag und Hausordnung." Visitor's Center, Gartenstadt Margarethenhöhe, Essen.

Markham, William T. 2005. "Networking Local Environmental Groups in Germany: The Rise and Fall of the Federal Alliance of Citizens' Initiatives for Environmental Protection (BBU)." *Environmental Politics* 14, no. 5:667–85.

McCarthy, Thomas. 1991. Introduction to *The Structural Transformation of the Public Sphere*, by Jürgen Habermas, trans. Thomas Burger, xi–xiv. Cambridge, MA: MIT Press.

McCreary, Eugene Charles. 1964. "Essen, 1860–1914: A Case Study of the Impact of Industrialization on German Community Life." PhD diss., Yale University.

———. 1968. "Social Welfare and Business: The Krupp Welfare Program, 1860–1914." *Business History Review* 41, no. 1:24–49.

McDonnell, Terence E. 2010. "Cultural Objects as Objects: Materiality, Urban Space, and the Interpretation of AIDS Campaigns in Accra, Ghana." *American Journal of Sociology* 115, no. 6:1800–1852.

McMullen, Shannon Crystal. 2007. "Post-Industrial Nature and Culture in the Ruhr District, Germany, 1989–1999." PhD diss., University of California, San Diego.

Mesecke, Andrea. 2010. "Stadtplanung von unten: Vom Protest zum Programm." In *Urbanität gestalten: Stadtbaukultur in Essen und im Ruhrgebiet, 1900 bis 2010*, 239–46. Göttingen: Museum Folkwang/Steidl.

Metzendorf, Georg. 1906. *Denkschrift über den Ausbau des Stiftungsgeländes*. Essen-Rüttenscheid: Margarethe Krupp-Stiftung für Wohnungsfürsorge.

Metzendorf, Rainer, and Achim Mikuscheit. 1997. *Margarethenhöhe: Experiment und Leitbild*. Essen: Margarethe Krupp-Stiftung für Wohnungsfürsorge.

Miller, Wallis. 1993. "IBA's 'Models for a City': Housing and the Image of Cold-War Berlin." *Journal of Architectural Education* 46, no. 4:202–16.

Mische, Ann. 2009. "Projects and Possibilities: Researching Futures in Action." *Sociological Forum* 24, no. 3:694–704.

———. 2014. "Measuring Futures in Action: Projective Grammars in the Rio+20 Debates." *Theory and Society* 43, nos. 3–4:437–64.

Mitchell, Don. 1995. "The End of Public Space? People's Park, Definitions of the Public, and Democracy." *Annals of the Association of American Geographers* 85, no. 1:108–33.

Mitscherlich, Alexander. 1965. *Die Unwirtlichkeit unserer Städte: Anstiftung zum Unfrieden*. Frankfurt a.M.: Suhrkamp.

Mitscherlich, Alexander, and Margarete Mitscherlich. 1975. *The Inability to Mourn: Principles of Collective Behavior*. New York: Grove.

Molotch, Harvey. 2003. *Where Stuff Comes From: How Toasters, Toilets, Cars, Computers, and Many Other Things Come to Be as They Are*. New York: Routledge.

Montero, Sergio. 2017. "Worlding Bogotá's Ciclovía: From Urban Experiment to International 'Best Practice.'" *Latin American Perspectives* 44, no. 2:111–31.

Moore, Jason W. 2015. *Capitalism in the Web of Life: Ecology and the Accumulation of Capital*. London: Verso.

Moses, A. Dirk. 1999. "The Forty-Fivers: A Generation between Fascism and Democracy." *German Politics and Society* 17, no. 1:94–126.

Mössner, Samuel, and Byron Miller. 2015. "Sustainability in One Place? Dilemmas of Sustainability Governance in the Freiburg Metropolitan Region." *Regions Magazine* 300, no. 1:18–20.

Mubanga, Mwenya, Liisa Byberg, Christoph Nowak, Agneta Egenvall, Patrik K. Magnusson, Erik Ingelsson, and Tove Fall. 2017. "Dog Ownership and the Risk of Cardiovascular Disease and Death—a Nationwide Cohort Study." *Scientific Reports* 7, no. 1:15821.

Mukerji, Chandra. 1997. *Territorial Ambitions and the Gardens of Versailles*. Cambridge: Cambridge University Press.

Museum Folkwang, ed. 2010. *Urbanität gestalten: Stadtbaukultur in Essen und im Ruhrgebiet, 1900 bis 2010*. Göttingen: Museum Folkwang/Steidl.

Nash, Roderick Frazier. 2014. *Wilderness and the American Mind*. New Haven, CT: Yale University Press.

Negt, Oskar. 1978. "Produzieren freie Schulen soziale Krüppel?" In *Jahrbuch für Lehrer 1978*, ed. Johannes Beck and Heiner Boehncke, 84–95. Reinbek bei Hamburg: Rowohlt Taschenbuch Verlag.

Negt, Oskar, and Alexander Kluge. 1972a. *Öffentlichkeit und Erfahrung: Zur Organisationsanalyse von bürgerlicher und proletarischer Öffentlichkeit*. Edition Suhrkamp, vol. 639. Frankfurt: Suhrkamp Verlag.

———. (1972b) 1993. *Public Sphere and Experience*. Translated by Peter Labanyi, Jamie Owen Daniel, and Assenka Oksiloff. Minneapolis: University of Minnesota Press.

Nelles, Wilfried, and Reinhard Oppermann. 1979. *Stadtsanierung und Bürgerbeteiligung*. Göttingen: Schwartz.

Nelson, Glenn. 2015. "Why Are Our Parks So White?" *New York Times*, July 10, SR4.

Newman, Andrew. 2015. *Landscape of Discontent: Urban Sustainability in Immigrant Paris*. Minneapolis: University of Minnesota Press.

Niethammer, Lutz. 1983. *Lebensgeschichte und Sozialkultur im Ruhrgebiet, 1930 bis 1960*. Bonn: J. H. W. Dietz.

Niklaß, Anja. 1999. *"Wenn die Gewaltigen klug sind": Die Essener Wohnungs- und Bodenpolitik, 1885–1915*. Marburg: Tectum.

Nolan, Mary. 2003. *Social Democracy and Society: Working Class Radicalism in Düsseldorf, 1890–1920*. Cambridge: Cambridge University Press.

Osthaus, Karl Ernst. 1911a. "Die Bedeutung der Gartenstadtbewegung für die künstlerische Entwickelung unserer Zeit." In *Die deutsche Gartenstadtbewegung: Zusammenfassende Darstellung über den heutigen Stand der Bewegung*, 99–101. Berlin. http://www.keom02.de/KEOM%202001/archive/dm/z100a.html.

———. 1911b. "Gartenstadt und Städtebau." In *Bauordnung und Bebauungsplan, ihre Bedeutung für die Gartenstadtbewegung*, ed. Deutsche Gartenstadt-Gesellschaft (Berlin-Schlachtensee), 33–40. Paris: Renaissance.

Park, B. J., Y. Tsunetsugu, T. Kasetani, T. Kagawa, and Y. Miyazaki. 2010. "The Physiological Effects of Shinrin-Yoku (Taking in the Forest Atmosphere or Forest Bathing): Evidence from Field Experiments in 24 Forests across Japan." *Environmental Health and Preventative Medicine* 15, no. 1:18–26.

Park, Lisa Sun-Hee, and David Naguib Pellow. 2011. *The Slums of Aspen: Immigrants vs. the Environment in America's Eden*. New York: New York University Press.

Park, Robert Ezra, and Ernest Watson Burgess. (1925) 1984. *The City*. Reprint, Chicago: University of Chicago Press.

Patel, Varsha. 2015. "Going Green? Washing Stones in World-Class Delhi." In *Sustainability in the Global City: Myth and Practice*, ed Cindy Isenhour, Gary McDonogh, and Melissa Checker, 82–105. Cambridge: Cambridge University Press.

Peck, Jamie, and Nik Theodore. 2015. *Fast Policy*. Minneapolis: University of Minnesota Press.

Peters, Ralf. 1999. *100 Jahre Wasserwirtschaft im Revier: Die Emschergenossenschaft, 1899–1999*. Bottrop: Pomp.

Pieper, L. 1903. *Die Lage der Bergarbeiter im Ruhrgebiet*. Stuttgart and Berlin: Studien.

Pizonka, Sonja. 2010. "Postindustrielle Grünplanung und ihre Grundlagen." In *Urbanität Gestalten: Stadtbaukultur in Essen und im Ruhrgebiet, 1900 bis 2010*, ed. Museum Folkwang Essen, 159–64. Göttingen: Museum Folkwang/Steidl.

Popitz, Heinrich, Hans Paul Bahrdt, Ernst August Jüres, and Hanno Kesting. 1957. *Das Gesellschaftsbild des Arbeiters: Soziologische Untersuchungen in der Hüttenindustrie.* Tübingen: Mohr.

Pounds, Norman J. G. 1968. *The Ruhr: A Study in Historical and Economic Geography.* New York: Greenwood.

Priebs, Axel. 2019. *Die Stadtregion: Planung—Politik—Management.* Stuttgart: Verlag Eugen Ulmer.

Projektgruppe Eisenheim mit Jörg Boström und Roland Günter. 1973. *Rettet Eisenheim.* Bielefeld: Verlag für das Studium der Arbeiterbewegungen.

Projekt Ruhr GmbH. 2010. *Masterplan Emscher Landschaftspark 2010.* Edited by Michael Schwarze-Rodrian. Essen: Klartext.

Prossek, Achim. 2004. "A Coal Mine Is Not a Coal Mine: Image Improvement and Symbolic Representation of the Ruhr Area, Germany." In *City Images and Urban Regeneration*, ed. Frank Eckart and Peter Kreisl, 67–81. Frankfurt: Peter Lang.

———. 2006. " 'Culture through Transformation—Transformation through Culture': Industrial Heritage in the Ruhr Region—the Example of Zeche Zollverein." In *Heritage and Media in Europe—Contributing towards Integration and Regional Development*, ed. Dieter Hassenpflug, Burkhardt Kolbmüller, and Sebastian Schröder-Esch, 239–48. Weimar: Bauhaus Universität.

———. 2009. *Bild-Raum Ruhrgebiet: Zur symbolischen Produktion der Region.* Detmold: Dorothea Rohn.

Provoost, Michelle, and Wouter Vanstiphout. 2011. "WiMBY! Welcome into My Backyard!: How to Revive an Area by Using What's Already There." In *Internationale Bauausstellung Emscher Park: Impulse*, ed. Christa Reicher, Lars Niemann, and Angela Uttke, 258–68. Essen: Klartext.

Prowe, Diethelm. 2001. "The 'Miracle' of the Political-Culture Shift: Democratization between Americanization and Conservative Reintegration." In *The Miracle Years*, ed. Hanna Schissler, 451–58. Princeton, NJ: Princeton University Press.

Purvis, Trevor, and Alan Hunt. 1993. "Discourse, Ideology, Discourse, Ideology, Discourse, Ideology. . . ." *British Journal of Sociology* 44, no. 3:473–99.

Rainey, John. 2014. "New York's High Line Park: An Example of Successful Economic Development." *Leading Edge* (Fall/Winter). https://www.greenplayllc.com/wp-content/uploads/2014/11/Highline.pdf.

Rehfeld, Dieter. 1995. "Disintegration and Reintegration of Production Clusters in the Ruhr Area." In *The Rise of the Rustbelt*, ed. Phil Cooke, 85–103. New York: St. Martin's.

Reicher, Christa, Klaus R. Kunzmann, Jan Polívka, Frank Roost, Yasemi Utku, and Michael Wegener. 2011. *Schichten einer Region.* Berlin: Jovis.

Reicher, Christa, Lars Niemann, and Angela Uttke, eds. 2011. *Internationale Bauausstellung Emscher Park: Impulse.* Essen: Klartext.

Reichl, Alexander. 2016. "The High Line and the Ideal of Democratic Public Space." *Urban Geography* 37, no. 6:904–25.

Reichow, Hans Bernhard. 1959. *Die autogerechte Stadt.* Ravensburg: Otto Maier.

Richter, Ralph. 2017. "Industrial Heritage in Urban Imaginaries and City Images." *Public Historian* 39, no. 4:1533–76.
Ritter, Gerhard A. 1978. "Workers' Culture in Imperial Germany: Problems and Points of Departure for Research." *Journal of Contemporary History* 13:165–89.
Robinson, Jennifer. 2002. "Global and World Cities: A View from Off the Map." *International Journal of Urban and Regional Research* 26, no. 3:531–54.
———. 2006. *Ordinary Cities*. London: Routledge.
Roseman, Mark. 1992. *Recasting the Ruhr, 1945–1958*. Oxford: Berg.
Rosenzweig, Roy, and Elizabeth Blackmar. 1992. *The Park and the People: A History of Central Park*. New York: Henry Holt.
Roth, Roland. 1991. "Local Green Politics in West German Cities." *International Journal of Urban and Regional Research* 15, no. 1:75–89.
Roy, Ananya. 2009. "The 21st-Century Metropolis: New Geographies of Theory." *Regional Studies* 43, no. 6:819–30.
Rubio, Fernando Domínguez. 2014. "Preserving the Unpreservable: Docile and Unruly Objects at MoMA." *Theory and Society* 43, no. 6:617–45.
RVR (Regionalverband Ruhr). 2014a. "Metropole Ruhr: Marketing for the Region." http://www.metroperuhr.de/en/home/the-ruhr-regional-association/public-relations.html (link inactive).
———. 2014b. "Metropole Ruhr: New Insights into the Ruhr Region." http://www.metroperuhr.de/en/start (link inactive).
Salin, Edgar. 1960. "Urbanität." In *Erneuerung unserer Städte: Vorträge, Aussprachen und Ergebnisse der 11. Hauptversammlung des deutschen Städtetages, Augsburg, 1.-3. Juni 1960*, 9–34. Stuttgart and Cologne: Neue Schriften des deutschen Städtetages.
———. 1970. "Von der Urbanität zur 'Urbanistik.'" *Kyklos* 23, no. 4:869–81.
Scammell, Margaret. 2000. "The Internet and Civic Enagement: The Age of the Citizen-Consumer." *Political Communication* 17, no. 4:351–55.
Schelsky, Helmut. 1963. *Die skeptische Generation*. Jena: Diederichs.
Schildt, Axel, and Alex Schildt. 1996. "From Reconstruction to 'Leisure Society': Free Time, Recreational Behaviour and the Discourse on Leisure Time in the West German Recovery Society of the 1950s." *Contemporary European History* 5, no. 2:191–222.
Schille, Peter, and Timm Rautert. 1973. "Heimat oder Hochhaus?" *Zeit Magazin* 23, no. 1 (June): 2–8.
Schilling, Joseph, and Jonathan Logan. 2008. "Greening the Rust Belt: A Green Infrastructure Model for Right Sizing America's Shrinking Cities." *Journal of the American Planning Association* 74, no. 4:451–66.
Schmid, Christian. 2014. "Travelling Warrior and Complete Urbanization in Switzerland: Landscape as Lived Space." In *Implosions/Explosions: Towards a Study of Planetary Urbanization*, 90–102. Berlin: Jovis.
Schmidt, Gert. 1969. "The Industrial Enterprise, History and Society: The Dilemma of German 'Industrie- und Betriebssoziologie.'" *Social Science Information* 8, no. 6:117–33.
Schmidt, Robert. 1912a. "Ein modernes Stadtgebilde: Die Industrie- und Wohnstadt." In *Essens Entwicklung, 1812–1912, herausgegeben aus Anlaß der hundertjährigen Jubelfeier der Firma Krupp*, 34–42. Essen: Fredebeul & Koenen.

———. (1912b) 2009. *Denkschrift betreffend Grundsätze zur Aufstellung eines General-Siedelungsplanes für den Regierungsbezirk Düsseldorf.* Reprint, Essen: Klartext.
Schmitt, Peter J. 1990. *Back to Nature: The Arcadian Myth in Urban America.* Baltimore: Johns Hopkins University Press.
Schofer, Lawrence. 1975. *The Formation of a Modern Labor Force: Upper Silesia, 1865–1914.* Berkeley: University of California Press.
Schubert, Dirk. 2004. "Theodor Fritsch and the German (*völkische*) Version of the Garden City: The Garden City Invented Two Years Before Ebenezer Howard." *Planning Perspectives* 19:3–35.
Schulte, Birgit. 2009. "Karl Ernst Osthaus, Folkwang and the 'Hagener Impuls': Transcending the Walls of the Museum." *Journal of the History of Collections* 21, no. 2:213–20.
Scott, James. 1998. *Seeing Like a State.* New Haven, CT: Yale University Press.
Scott, Joan Wallach. 1974. *The Glassworkers of Carmaux: French Craftsmen and Political Action in a Nineteenth-Century City.* Cambridge, MA: Harvard University Press.
Sennett, Richard. 1977. *The Fall of Public Man.* New York: Norton.
———. 1992. *The Conscience of the Eye: The Design and Social Life of Cities.* New York: Norton.
———. 1996. *Flesh and Stone: The Body and the City in Western Civilization.* New York: Norton.
Sevilla-Buitrago, Alvaro. 2017. "Gramsci and Foucault in Central Park: Environmental Hegemonies, Pedagogical Spaces and Integral State Formations." *Environment and Planning D: Society and Space* 35, no. 1:165–83.
Sewell, William, Jr. 1996. "Historical Events as Transformations of Structures: Inventing Revolution at the Bastille." *Theory and Society* 25, no. 6:841–81.
Shabecoff, Philip. 1965. "Germany's Ruhr in State of Flux." *New York Times*, May 10, 51–53.
Shay, Alice. 2012. "The Contemporary International Building Exhibition (IBA): Innovative Regeneration Strategies in Germany." PhD diss., Massachusetts Institute of Technology.
Shiner, Helen. 1997. "Artistic Radicalism and Radical Conservatism: Moïssy Kogan and His German Patrons, 1903–1928." PhD diss., University of Central England.
Siebel, Walter. 1999. "Industrial Past and Urban Future in the Ruhr." In *Cities in Transition*, ed. Randall Smith and Bernhard Blanke, 123–34. New York: Palgrave Macmillan.
Siegfried, Detlef. 2005. " 'Don't Trust Anyone Older Than 30?' Voices of Conflict and Consensus between Generations in 1960s West Germany." *Journal of Contemporary History* 40, no. 4:727–44.
Sieverts, Thomas. [1997] 2003. *Cities without Cities: An Interpretation of the Zwischenstadt.* Translated by Thomas Sieverts and Hildebrand Frey. New York: Routledge.
Simmel, Georg. (1902) 1964. "The Metropolis and Mental Life." In *The Sociology of Georg Simmel*, ed. K. H. Wolff, 409–24. New York: Free Press.
Sinclair, Upton. (1906) 2006. *The Jungle.* New York: Modern Library.
Sitte, Camillo. 1889. *Der Städtebau nach seinen künstlerischen Grundsätzen* (City planning according to artistic principles). Vienna: Graeser.
Slach, Ondrej, Petr Rumpel, and Tomas Boruta. 2011. "Transferable Impulses of IBA Emscher Park." In *Internationale Bauausstellung Emscher Park: Impulse*, ed. Christa Reicher, Lars Niemann, and Angela Uttke, 210–23. Essen: Klartext.

Smith, Gordon. 1976. "West Germany and the Politics of Centrality." *Government and Opposition* 11, no. 4:387–407.
Smith, Neil. (1984) 2010. *Uneven Development: Nature, Capital, and the Production of Space*. Reprint, Athens: University of Georgia Press.
———. 1996. *The New Urban Frontier: Gentrification and the Revanchist City*. New York: Routledge.
———. 1998. "Nature at the Millennium: Production and Re-Enchantment." In *Remaking Reality: Nature at the Millennium*, ed. Bruce Braun and Noel Castree, 271–85. New York: Routledge.
Smith, Suzanne. 2011. "The Institutional and Intellectual Origins of Rural Sociology." Paper presented at the 74th annual meeting of the Rural Sociology Society, July 28–31, Boise.
Soeffner, Hans-Georg. 1997. *The Order of Rituals*. New Brunswick, NJ: Transaction.
Somers, Margaret. 2008. *Genealogies of Citizenship: Markets, Statelessness, and the Right to Have Rights*. Cambridge: Cambridge University Press.
Spelt, Jacob. 1969. "The Ruhr and Its Coal Industry in the Middle 60s." *Canadian Geographer* 13, no. 1:3–9.
Sperber, Jonathan. 1997. "Bürger, Bürgertum, Bürgerlichkeit, Bürgerliche Gesellschaft: Studies of the German (Upper) Middle Class and Its Sociocultural World." *Journal of Modern History* 69, no. 2:271–97.
Stadt Dortmund. 2012a. "Evaluationsbericht über die kleinräumige Quartiersanalyse 'Hörder Neumarkt.'" https://www.dortmund.de/media/p/stadterneuerung/stadterneuerung_down loads/quartiersanalyse/Evaluationsbericht_Hoerder_Neumarkt.pdf.
———. 2012b. "Kurz- und Abschlussbericht: Kleinräumige Quartiersanalyse 'Hörde-Phoenix See.'" https://www.dortmund.de/media/p/stadterneuerung/stadterneuerung_downloads /quartiersanalyse/Abschlussbericht_Hoerde-Phoenix_See.pdf.
Stadt Dortmund—Dortmund Agenteur. 2015. "Economy and Science." http://www.dortmund .de/en/economy_and_science/home_es/index.html.
Stahlberg, Gertrude. 1957. *Die Vertriebenen in Nordrhein-Westfalen*. Berlin: Duncker & Humblot.
Steinborn, Vera. 1991. *Arbeitergärten im Ruhrgebiet*. Recklinghausen: Westfälisches Industriemuseum (Landschaftsverband Westfalen-Lippe).
———. 2010. "Arbeitergärten im Ruhrgebiet." In *Zwischen Kappes und Zypressen: Gartenkunst an Emscher und Ruhr*, ed. Martina Oldengott and Christine Vogt, 52–58. Essen: Klartext.
Steinhauer, Gerhard. 1956. *Gartenstadt Margarethenhöhe*. Essen: Margarethe Krupp-Stiftung für Wohnungsfürsorge.
Steinmetz, George. 2005. "The Epistemological Unconscious of US Sociology and the Transition to Post-Fordism: The Case of Historical Sociology." In *Remaking Modernity: Politics, History, and Sociology*, ed. Julia Adams, Elisabeth S. Clemens, and Ann Shola Orloff, 109–57. Durham, NC: Duke University Press.
Stonge, Carmen Luise. 1993. "Karl Ernst Osthaus: The Folkwang Museum and the Dissemination of International Modernism." PhD diss., City University of New York.
Strand, Michael. 2015. "The Genesis and Structure of Moral Universalism: Social Justice in Victorian Britain, 1834–1901." *Theory and Society* 44, no. 6:537–73.

Sum, Ngai-Ling, and Bob Jessop. 2013. *Towards a Cultural Political Economy: Putting Culture in Its Place in Political Economy.* Northampton: Edward Elgar.

SVR (Siedlungsverband Ruhrkohlenbezirk). [1960/61]. *Tätigkeitsbericht, 1958–60.* Essen: Verbandsdirecktor des Siedlungsverbandes Ruhrkohlenbezirk, Essen.

———. 1966. *SVR Gebietsentwicklungsplan, 1966.* Essen: Deutscher Gemeindeverlag, W. Kohlhammer.

———. 1969. *Industriestandort Ruhr.* Essen: Siedlungsverband Ruhrkohlenbezirk.

———. 1974. *Grüne Halden im Ruhrgebiet/Green Colliery Spoil Banks in the Ruhr/Terrils miniers amenagées en espaces verts dans la Ruhr.* Essen: Siedlungsverband Ruhrkohlenbezirk.

———. 1977. *Hier bin ich Mensch: Oasen einer Industrielandschaft: Die Revierparks im Ruhrgebiet.* Gelsenkirchen: Druckhaus Louisgang.

Swidler, Ann. 1986. "Culture in Action: Symbols and Strategies." *American Sociological Review* 51, no. 2:273–86.

Sze, Julie. 2015. *Fantasy Islands: Chinese Dreams and Ecological Fears in an Age of Climate Crisis.* Berkeley and Los Angeles: University of California Press.

Tate, Alan. 2013. *Great City Parks.* New York: Taylor & Francis.

Tavory, Iddo. 2011. "The Question of Moral Action: A Formalist Position." *Sociological Theory* 29, no. 4:272–93.

Tavory, Iddo, and Nina Eliasoph. 2013. "Coordinating Futures: Toward a Theory of Anticipation." *American Journal of Sociology* 118, no. 4:908–42.

Tavory, Iddo, and Colin Jerolmack. 2014. "Molds and Totems: Nonhumans and the Constitution of the Social Self." *Sociological Theory* 32, no. 1:64–77.

Taylor, Charles. 2004. *Modern Social Imaginaries.* Durham, NC: Duke University Press.

Taylor, Dorceta. 2016. *The Rise of the American Conservation Movement: Power, Privilege, and Environmental Protection.* Durham, NC: Duke University Press.

Tenfelde, Klaus. 2000. "Neue Mitte, neues Selbstbewusstsein." In *Ruhrstadt: Die Andere Metropole*, ed. Gerd Williamowski, Dieter Nellen, and Manfred Bourrée, 16–20. Essen: Klartext.

———, ed. 2005. *Krupp—the Rise of a World-Class German Company.* London: Philip Wilson.

Tenfelde, Klaus, and Thomas Urban, eds. 2010. *Das Ruhrgebiet: Ein historisches Lesebuch.* Essen: Klartext.

Thomas, Nick. 2003. *Protest Movements in 1960s West Germany: A Social History of Dissent and Democracy.* Oxford: Berg.

Thompson, E. P. (1963) 2016. *The Making of the English Working Class.* New York: Open Road Media.

Tönnies, Ferdinand. [1887] 2011. *Community and Society.* Reprint, Mineola, NY: Dover.

Treiber, Hubert, and Heinz Steinert. 1980. *Die Fabrikation des zuverlässigen Menschen.* Munich: Westfälisches Dampfboot.

Tsing, Anna Lowenhaupt. 2015. *The Mushroom at the End of the World: On the Possibility of Life in Capitalist Ruins.* Princeton, NJ: Princeton University Press.

Tumber, Catherine. 2011. *Small, Gritty, and Green: The Promise of America's Smaller Industrial Cities in a Low-Carbon World.* Cambridge, MA: MIT Press.

Ulrich, Roger S. 1993. "Biophilia, Biophobia, and Natural Landscapes." In *The Biophilia Hypothesis*, ed Stephen R. Kellert and Edward O. Wilson, 73–137. Washington, DC: Island.

———. 2002. "Health Benefits of Gardens in Hospitals." Paper presented at the conference "Plants for People International Exhibition Floriade," Haarlemmermeer, Netherlands. https://plantsolutions.com/documents/HealthSettingsUlrich.pdf.

Umbach, Maiken. 2009. *German Cities and Bourgeois Modernism, 1890–1924*. New York: Oxford University Press.

UN-Habitat. UN Human Settlement Programme. 2007. *The State of the World's Cities, 2006/2007: 30 Years of Shaping the Habitat Agenda*. London: Earthscan.

Van Zee, Marynel Ryan. 2012. "Form and Reform: The Garden City of Hellerau-bei-Dresden, Germany, between Company Town and Model Town." In *Company Towns*, ed. Marcelo J. Borges, and Susana B. Torres, 41–67. New York: Palgrave Macmillan.

Vertovec, Steven. 2012. "'Diversity' and the Social Imaginary." *European Journal of Sociology* 53, no. 3:287–312.

von der Wielbecke, Johann. 1965. "Marler Bergleute um die Jahrhundertwende." In *Vestischer Kalender*, vol. 37.

von Einem, Eberhard. 1982. "National Urban Policy—the Case of West Germany." *Journal of the American Planning Association* 48, no. 1:9–23.

Von Hodenberg, Christina. 2006. "Mass Media and the Generation of Conflict: West Germany's Long Sixties and the Formation of a Critical Public Sphere." *Contemporary European History* 15, no. 3:367–95.

Von Merveldt, Dieter. 1971. *Grossstädtische Kommunikationsmuster*. Cologne: Bachem.

Von Petz, Ursula. 1990. "Margarethenhöhe Essen: Garden City, Workers' Colony or Satellite Town?" *Planning History* 12, no. 2:3–9.

———. 1999. "Robert Schmidt and the Public Park Policy in the Ruhr District, 1900–1930." *Planning Perspectives* 14, no. 2:163–82.

Vormann, Boris. 2015. "Toward an Infastructural Critique of Urban Change: Obsolescence and Changing Perceptions of New York City's Waterfront." *City* 19, nos. 2–3:356–64.

Wachsmuth, David. 2012. "Three Ecologies: Urban Metabolism and the Society-Nature Opposition." *Sociological Quarterly* 53, no. 4:506–23.

Wachsmuth, David, and Hillary Angelo. 2018. "Green and Grey: New Ideologies of Nature in Urban Sustainability Policy." *Annals of the Association of American Geographers* 108, no. 4:1038–56.

Wachsmuth, David, Daniel Aldana Cohen, and Hillary Angelo. 2016. "Expand the Frontiers of Urban Sustainability." *Nature News* 536, no. 7617:391–93.

Wachten, Kunibert, ed. 1996. *Wandel ohne Wachstum? Change without Growth? Extended Catalogue of the Official German Presentation at the VI. Architecture Biennale Venice, 1996*. Braunschweig and Wiesbaden: Friedr. Vieweg & Sohn.

Walton, John. 1992. *Western Times and Water Wars*. Berkeley and Los Angeles: University of California Press.

Warner, Michael. 2002. "Publics and Counterpublics." *Public Culture* 14, no. 1:49–90.

Weber, Annemarie. 1957. "Für die Freizeit gerüstet?" *Rheinischer Merkur*, October 11.

Weber, Eric. 1963. *Das Freizeitproblem: Anthropologisch-pädagogische Untersuchung*. Munich: Ernst Reinhardt.

Weilacher, Udo. 2010. "Ferme Ornée Mechtenberg. Field Experiments between Post-Industrial Wilderness and New Usefulness." In *Feldstudien*, 84–89. Basel: Birkhäuser.
Weiss, Joachim, Wolfgang Burghardt, Peter Gausmann, Rita Haag, Henning Haeupler, Michael Hamman, Bertram Leder, Annette Schulte, and Ingrid Stempelmann. 2005. "Nature Returns to Abandoned Industrial Land: Monitoring Succession in Urban-Industrial Woodlands in the German Ruhr." In *Wild Urban Woodlands*, ed. Ingo Kowarik and Stefan Körner, 143–62. Berlin and Heidelberg: Springer.
White, Monica. 2011. "Sisters of the Soil: Urban Gardening as Resistance in Detroit." *Race/Ethnicity: Multidisciplinary Global Contexts* 5, no. 1:13–28.
Whyte, William H. 1956. *The Organization Man*. New York: Simon & Schuster.
——. 1980. *The Social Life of Small Urban Spaces*. New York: Project for Public Spaces.
Wild, Martin Trevor. 1983. *Urban and Rural Change in West Germany*. Lanham, MD: Rowman & Littlefield.
Williams, Raymond. 1958. *Culture and Society, 1780–1950*. London: Chatto & Windus.
——. 1973. *The Country and the City*. Oxford: Oxford University Press.
——. 1977. *Marxism and Literature*. New York: Oxford University Press.
——. 1995. *The Sociology of Culture*. Chicago: University of Chicago Press.
——. 2005. *Culture and Materialism*. New York: Verso.
Winner, Langdon. 1980. "Do Artifacts Have Politics?" *Daedalus* 109, no. 1:121–36.
Wirth, Louis. 1938. "Urbanism as a Way of Life." *American Journal of Sociology* 44, no. 1:1–24.
Wischermann, Clemens. 1986. "Wohnungsmarkt, Wohnungsversorgung, und Wohnmobilität in deutschen Großstädten, 1870–1913." In *Stadtwachstum, Industrialisierung, Sozialer Wandel*, ed. Peter Borscheid, Bruno Fritzsche, Friedrich-Wilhelm Henning, Dietmar Petzina, Günther Schulz, Hans-Jürgen Teuteberg, Richard H. Tilly, and Clemens Wischermann, 101–33. Berlin: Duncker & Humblot.
Wuppertal Institut für Klima, Umwelt, Energie. 2013. *Emscher 3.0*. Bönen: Kettler.
Young, Iris Marion. 1990. *Justice and the Politics of Difference*. Princeton, NJ: Princeton University Press.
Zöpel, Christoph. 2010. "Die Metropole Ruhr—zur Rolle Dortmunds." In *Stadtentwicklung in Dortmund seit 1945*, ed. H. Bömer, E. Lürig, Y. Utku, and D. Zimmerman, 47–60. Dortmund: Institut für Raumplanung, Technische Universität Dortmund.
"Zu blauen Himmeln." 1961. *Der Spiegel*, no. 33 (September 8): 22–33.
Zukin, Sharon, and Jennifer Smith Maguire. 2004. "Consumers and Consumption." *Annual Review of Sociology* 30:173–97.

# INDEX

Abu Dhabi (United Arab Emirates), 2, 6, 207
Addams, Jane, 35
agora, 80–81
agrarian community: romantic nostalgia for, 54; social control, 40–41
Alberta, 16
allotment gardens (*Schrebergärten*), 39–40, 128–30
*All That Is Solid Melts into Air* (Berman), 103
American West, 29
ancien régime, 81–82
Anderson, Benedict, 18–19, 73, 207; comparative field, 52
animals, 62, 109–10, 128, 166, 186, 210; access to, 40, 43; benefits of, 3; in colonies, 38–40, 43, 56–57, 61, 121–22, 126–27, 129–30, 136; Eisenheim movement, 121–22, 125–27; forbidding of, 56; as goods, 3, 17, 57, 129; in Margarethenhöhe, 56–57; in Ruhr Valley, 48, 101, 130; social and symbolic purposes, 130; subsistence purposes, 17, 39, 57, 129; urbanized nature, 15, 57
antigreening arguments, 216
Asia, 17, 203, 206
Athens Charter, 84, 89
August Thyssen Hütte (ATH), 107, 132

Bahrdt, Hans-Paul, 84, 105, 117, 120
Bauhaus, 44
Becher, Bernd, 136–37
Becher, Hilla, 136–37
Berlin (Germany), 31, 34–36, 43–49, 66, 69, 73, 86, 89, 173–74; Brandenburg Gate, 159, 165, 185
Berlin Wall, 143
Berman, Marshall, 103
Bielefeld (Germany), 87–88
Birmingham school, 116
Bloomberg, Michael, 212–13
Bogotá (Colombia), 148
Bonn (Germany), 85
*BOS Steelmaking Shop* (Hebig), 167
Bottrop (Germany), Tetrahedron, 161–62
bourgeoisie, 116, 121, 133, 203, 210; cultural space, 209; leisure, 129; politics, 120, 128; publics, 202; public sphere, 78, 80–81, 94, 100, 108, 117–20, 124, 135–36, 208; rationality, 108; urbanized nature, 129
Brandt, Willy, 86, 102, 131; "blue skies" speech, 101, 185
Brasília (Brazil), 84
Brownstone Detectives, 12

*Bürgers*, 64
Burgess, Ernest W., 7

capitalism, 16, 105; industrial, countryside, longing for, 29
Castoriadis, Cornelius, 7, 20, 112, 118, 134, 207
Cézanne, Paul, 46
Chicago (Illinois), 29, 35, 44, 58, 160
Chicago school, 7
China: green urbanism, 2; and industrialization, 148–49, 172
cholera, 67
Christian Democratic Union, 101–2, 132
cities, 16; as aspirational, 43, 47; as category of practice, 15; centrality, 34; citymaking, green-as-good logic, 203; density, 34; greening of, 23, 57; heterogeneity, 34; as "highly imageable environments," 160; improvement, through signifiers of nature, 19, 57; industrial cities, romantic view of nature, 29; and nature, 6–8, 201, 204; as social and spatial ideal, 30; trees in, 216; urbanized nature, 20. *See also* green cities
*Cities without Cities* (Sieverts), 154–55
citizenship: belonging, 215; green space, 215; ideals of, 147; as practice of consumption, 151; race and class, 215; urbanized nature, 73
city beautiful movement, 13
*City Planning according to Artistic Principles* (Sitte), 55
climate change, 1, 204–5, 217–18
coal crisis, 85–86, 117
Cold War, 212
Cologne (Germany), 198–99; Dome, 165, 185
colonialism, 16
colonies, 25, 30–31, 37, 55, 70–72, 78, 81–82, 94–95, 98, 132, 214; animals in, 38–40, 43, 56–57, 61, 121–22, 126–27, 129–30, 136; benefits of, 110; children, socialization of, 127; class consciousness, 131; community, construction of, 121–22; counterpublic sphere, fostering of, 108, 121; deindustrialization, 131, 133; demolishing of, 107; employers, as benevolent patrons, 41–42; ethnic and linguistic enclaves, 39; gardening, 42; gardens, 43, 60, 121, 125–30; green rooms, 122, 124; green space, 39, 60, 109–10, 121, 128–29; half-public sphere, 122–23; housing shortage, 48; in-between areas, 122; industrial workers, as peasant-farmers, 41–42; integration, governing principle of, 122; leisure education, 127; leisure time, 126–27; local activism, 42; material continuity, 129; nature, access to, 39–41, 110, 113, 131; nature, reimagined as, 113; nature, subsistence purposes of, 56; neighborly relations, 124; paternalism, feudal form of, 41, 69; political conservatism, association with, 109; proletarian politics, as haven for, 113, 121; reciprocity, 128; reinvention of, 111, 113, 131; residents, grouping of, 62; rural imaginations, 39; rural life, facsimiles of, 43; solidarity, 128, 134; as village-like, 38–40; working-class life, as symbols of, 111. *See also* company housing
common good, 106
community gardens, 217
company housing, 65, 71, 95, 109, 111, 132; agricultural life, mimicking of, 37; as desirable and emancipatory space, 133; garden plot, 40; good life, 40; proletarian politics, as haven for, 113; social control functions, 41; turnover rate, 41; worker retention, solution to, 41. *See also* colonies
competitiveness, 17, 155–56, 177, 180, 185, 194; comparative field, 52, 57; planning process, 200; regional, 9; territorial, 150
conservation, 29
cosmopolis, 81–82
cosmopolitanism, 19, 62–64
counterpublic sphere: green rooms, 124; as proletarian, 108–9, 121. *See also* public sphere

*Country and the City, The* (Williams), 12–13
cultural creativity, 112
culture, materiality of, 202

deindustrialization, 13, 86–87, 113–14, 131, 133, 159, 161, 186
democracy, 103, 111, 134, 214; abstract framework for, 80; anonymity and exposure to difference, 92; crisis of, 25; deliberative, 104, 106; democratic public life, 79–80; and democratization, 83, 114, 132, 209; green space, 78–80, 104, 215; leisure time, 96, 100, 104, 113; parks, access to, 92, 96, 105, 209; and participation, 132; planners' vision of, 89; public sphere, 79–80, 83, 88–89, 98, 106, 153; rebuilding of, 11, 77–78, 80; and *Revierparks*, 92, 95, 103–4, 116; space-times of, 79–80; as urban, 25, 78–80, 104, 108
desert cities, greenness of, 207
Detroit (Michigan), 10, 141, 148–49
development studies, 7
Dickens, Charles, 35
displacement, 194, 198, 204–5, 209, 216
Dortmund (Germany), 10, 14, 25–26, 32, 103, 137, 148–49, 163, 166, 176–79, 188; displacement, 194, 198; gentrification, 194–95, 198; high-tech imaginary, emphasis on, 181–82; quality of life, emphasis on, 184–86; repackaging of, 185; structural change, approach to, 185–86, 198; Turkish immigrants, 197–98, 200. *See also* Phoenix Lake
Dubai (United Arab Emirates), as green city, 203
Duisburg (Germany), 32, 135, 149, 158, 176
Durkheim, Émile, 7–8, 130, 165
Düsseldorf (Germany), 42–43, 87–88

East Germany, 141, 206. *See also* Germany; West Germany
ecology, 6–7, 114, 142, 154, 157
Eisenheim movement, 117–18, 135, 147, 153, 160; animals, 121, 122, 125–27; collaborative-solidary social structure, destruction of, 133; counterpublic sphere, advocating for, 108; and deindustrialization, 113, 134; democratic politics, critique of, 113; familiarity, 108, 123–24, 126; free time, as threat to democratic politics, 126; green rooms, 122–23; green space, 112–13, 131; green space, benefits of, 109; high-rise apartment block, targeting of, 113–14, 116; informality, 123–24; Left movement, 111; leisure activities, 127; leisure spaces, as political, 133–34; mutuality, 108, 126; and nature, 131, 134; neighborliness, 125–26; paternalistic aspect of, 134, 136; as playground, 126–28; preservation, 121; as proletarian, 109, 133–34; reciprocity, 126; reimagining, exploring of, 109; social qualities of, 124; solidarity, 108, 126, 133; spatial morphology of, 124; spontaneity, 123–24; urban space, 119; worker-specific discourse, 124; working class, depoliticization of, 120. *See also* colonies
Eisenheim workers' colony, 107–8, 132, 196, 208; campaign to save, 109; as idyll, 110; nature, access to, 110–11; preservation of, 136; as social and political space, 121; Turkish immigrants, 128; as working class, 113, 120, 131, 134–36
Emscher River, 31–32, 99, 100, 192; periodic floods, 48; renaturalization of, 157, 165, 182–83; restoration of, 142
Engels, Friedrich, 10, 34–35
England, 29, 47, 86, 117
environmental movements, 29, 215, 217
Essen (Germany), 10, 31, 33–34, 36, 39, 44, 47, 54, 59–61, 65, 107, 115, 132–33, 149, 164, 176; as European Capital of Culture, 9, 142, 156; greening in, 66–68; as leisure park, 45–46; Nordpark, 67–68; Stadtgarten, 67–68; Volksgarten, 66–67
Ethiopia, 204
Europe, 4, 9, 20, 30–31, 33–34, 43–44, 46, 51, 54, 58, 65, 72, 87, 143, 158–59, 184, 205; cosmopolis, 81–82; Greenest Block

Europe (*cont.*)
  contests, 2; greening projects, 17; nature, romantic view of, 29; postwar boom, 83; "shrinking" cities in, 203; social movements in, 114; spatial planning, regionalization of, 156; urbanity in, 10; urban public spaces, 81
European Union (EU), 143-44, 176, 189
Evert, Jürgen, 194-98

*Fall of Public Man, The* (Sennett), 81
*Faust* (Goethe), 93, 103-4, 126
Federal Building Act, 84
Federal Republic. *See* Germany
feudalism, 41-42
FitzSimmons, Margaret, 54, 56
*flânerie*, 66-67
Folkwang Museum, 46
Fordism, 5, 23, 77, 83, 103, 150-51
'45ers, 78, 94, 96, 103-5, 113-14, 116, 134
France, 11, 29, 86, 205-6
French Revolution, 205-6
Friedrich, Caspar David, 11
Friedrich Krupp AG, 71, 74; company housing, 34, 37-40; company stores, 64; growth of, 33-34; innovations of, 34, 44, 47; paternalistic labor model of, 37; political power of, 43-44; socialization goals, 63. *See also* Thyssenkrupp
Fritsch, Theodor, 55, 73
frontier, 30

Gans, Herbert, 120
Ganser, Karl, 145-47, 152, 154-55, 157, 164, 169
garden cities, 13, 30, 48, 54, 56, 61, 80, 82, 92, 121; anti-Semitic version of, 55; bourgeois subject formation, site of, 73; citizenship, ideals of, 147; cosmopolitan culture, expression of, 50-51; as fairy-tale German village, 55; mind, elevating of, 50; nature, as decorative and symbolic, 57; romantic nostalgia, 50; as spatial form, 50; town and country, fusion of, 47, 49, 59; urbanism, as model of, 49-50; urbanized nature, 31, 51-53, 72; urban life, achieving of, 59; urban vision of, 55
Garden City Association, 47
garden city movement, 48, 50
gardening, 42; associational life, 126; as hobby, 57
gardens, 39-40, 43, 60-61; as extension of home, 62; as leisure spaces, 71, 130; as retreats, 130; self-sufficiency, illusion of, 42; social life, organization of, 130-31; as spaces of relaxation, 57-58; spiritual needs, fulfilling of, 62; subsistence role of, 130
*Gemeinschaft* (community), 7, 24-25, 38, 97
General City Planning Exhibition, 44
gentrification, 194-96, 198, 216
German Garden City Association, 44, 47, 55-56
Germany, 4, 9-10, 31, 33-34, 43-44, 46, 55-56, 86, 93, 95, 105, 115, 149, 159, 164, 184, 192, 205-6; mass motorization and highway construction, 91; as nation-state, 35; nature and nationalism, 11; people's parks in, 79; planning exhibitions, 143; postwar boom, 83; romanticism in, 11; spatial planning, regionalization of, 156; urban life, decline in, 106; urban planning in, 85. *See also* East Germany; garden cities; Ruhr Valley; West Germany
*Gesellschaft* (society), 7, 24-25, 38, 40
gigantism, 60
Glasgow (Scotland), 173
globalization, 148-49, 153
Global South, 2
Goethe, Johann Wolfgang von, 93
Gramsci, Antonio, 71
green, 153; aesthetics, 217; association with good, 19, 57; social projects, as vehicle for, 211; urban redevelopment, tool for, 150-51
green cities, 1-4, 44, 54, 203; as entrepreneurial, 13

Greenest Block contest, 1–2, 19
greening, 1, 12, 17, 30, 51, 109, 147–48, 151, 170, 218; as aspirational, 111, 146; benevolent gifts, 208; of cities, 4, 6, 57, 112; as contemporary, 3; critique of, 216; gentrification, 216; global spread of, 3, 206; good intentions, 74, 212; as "grammar" of moral action, 5, 22; green buildings, 3, 204; green infrastructure, 204; green landscapes, as nature, 143; green planning, 204; as habit, 146; industrial cities, reaction to, 6; and leisure, 68; liberalism, compatibility with, 104; managerial aspects, 70–72, 74; materialist approach, 7; moral expression, portable form of, 111–12; municipal support for, 66–68; nation, idea of, 19; as paradoxical, 23; paternalism, form of, 24, 74, 212; as philanthropic practice, 71; as public good, 69, 71, 208; public politics, warping of, 217; as reinvention, 153; remaking cities, mode of, 5, 22–23; of Ruhr Valley, 65–66, 70, 72, 74, 77–78, 101, 141, 193, 202–7; signifying system, 13; social action, 206–7; social engineering, as form of, 69; social imaginaries, 20, 65; as social-managerial technology, 104; social organization, form of, 202; as social practice, 72–74, 152, 201; social relations, embedded in, 23; as spatial ordering, 89; status quo, upholding of, 71; streetscaping, green forms of, 3; universal benefits of, 25–26; as urban amenity, 71; urban change, as cause and effect, 20; urban greening, 141; urbanization, 206; urbanized nature, 71, 111–12, 207–8; urban publics, spatializing of, 81; white men, 208. *See also* urban greening
greening movements, 134
greening practices. *See* greening projects
greening projects, 22, 59, 74, 78, 80, 144, 153, 169, 174, 185, 207, 210–11; as aspirational, 145; as consumer good, 188; critique of, 193–97; displacement, 194; heterogeneity of, 136; as imaginative, 147; and leisure, 23–24; under neoliberalism, 199; paternalistic nature of, 170–71; public and private spheres, reconfiguring of, 81; as public goods, 23, 171, 187–88, 193–94, 200; in public politics, 21; in Ruhr Valley, 4–5, 9, 12–13, 20–21, 23, 30–31, 51, 148, 151; as sites of action, 13–14; as social projects, 212; as universal, 199–200, 212; universality, falling short, 208; and urbanization, 17; as urbanized nature, 15, 24, 143; as well-intentioned, 187, 193
green space, 1, 3, 12, 45, 59–62, 67, 71–73, 89, 92, 108, 110, 112–13, 129, 131, 171, 193, 199, 206, 212–13, 216–17; in cities, 2; citizenship, 215; in colonies, 39, 121; cosmopolitan culture, enhancing of, 68; decorative forms of, 65; and democracy, 78–80, 104, 215; Eisenheim movement, 109; as morally and socially beneficial, 82; policy implications of, 215; public ambitions, 82; as public realm, 66, 91; public sphere, 82; in Ruhr Valley, 24, 31, 50, 68–69, 78, 104, 111, 128, 130; social imaginaries, 78; spatiotemporal qualities of, 78; subsistence purposes, 39, 56, 130; as urban-aspirational, 56; urban democracy, 104; urban elite, 209; as urban form, 68; as urban infrastructure, 70; urban life, 8–9; urban public, 94; urban publicness, as organizing principle of, 203; urban public space, 81. *See also* trees; urban green space
green tourism, 161
green urbanism, 26; as policy transfer, 207
greenwashing, 5
green widows, 115
guest workers, 98, 128; in Silesia, 33; in Turkey, 98; in Yugoslavia, 98
Günter, Janne, 108, 119, 126–27
Günter, Roland, 14, 108, 126, 130, 133–36, 196, 208

Habermas, Jürgen, 14, 25, 92, 96, 104, 108, 113–14, 120, 125, 135, 154; communicative

Habermas, Jürgen (*cont.*)
  rationality, 98; critics of, 106, 114, 117–18; democratic vision of, 89; middle class, vision of, 100; public planners, influence on, 81; public sphere, 78, 80, 85, 94, 117–19, 124
Hall, Peter, 10
Hanover (Germany), 173–74
Hattingen (Germany), 32
Haussmann, Georges-Eugène, 202
Heatherwick, Thomas, 214
Hebig, Haiko, 167, 169
*Heimatschutz* movement. *See* nationalism
Herlyn, Ulfert, 105
high-rises, 84, 94, 107–10, 116, 120–24, 126–27, 129, 132; critique of, 115; and decay, 91, 115; defense of, 105; green windows, 115; isolation, as symbol of, 91; social mobility, 113–14; targeting of, 113; urban planning, 105
Hitler, Adolf, 91
*Hochofen Phoenix 3 (No. 3 Phoenix blast furnace)* (Hebig), 167
housing: crisis of, 34–37; and industrialization, 35; and paternalism, 37; shortages, 34–35, 48; and urbanism, 34
Howard, Ebenezer, 47, 49, 55, 59

*Image of the City, The* (Lynch), 159–60
*Imaginary Institution of Society, The* (Castoriadis), 20
*Imagined Communities* (Anderson), 18–19
indigenous peoples, 30
individualism, 64, 69, 116
industrialism, 41–42
industrialization, 7, 13, 33–34, 36, 46–47, 180; housing crisis, product of, 35; industrial brownfields, transformation of, 150; romanticized view of, 186; urbanization, 20, 30, 206
*Industriekultur* (industrial culture), 175–76, 183–84
*Industrienatur* (industrial nature), 154, 157–58, 160, 165–66, 173, 175, 177, 179–80, 182, 185

*Inhospitality of Our Cities, The* (Mitscherlich), 115
Internationale Bauausstellung (IBA) Emscher Park, 9, 14, 25–26, 141, 143, 146–48, 151–52, 156, 164, 167, 170, 172, 174, 188–90, 194–95, 197, 204; Duisburg-Nord, 158, 161–62, 173, 176–77, 179, 182, 214; Emscher Park Master Plan, 142; as imaginative project, 144–45; *Industriekultur* (industrial culture), 175–76, 183–84; *Industrienatur* (industrial nature), 154, 157–58, 160, 165–66, 173, 175, 177, 179, 180, 182, 185; out-of-everydayness, creating of, 165; as public good, 153; and recycling, 154, 157–59, 181, 186; soft management, 163; structural change, 150; *Zwischenstadt* (in-between city), 154–55, 160, 169, 177
Internationale Essener Songtage, 115
Italy, 33, 98, 160

Jacobs, Jane, 89, 120
Japan, 211
Jencks, Charles, 115
*Jungle, The* (Sinclair), 58–59, 71

Kluge, Alexander, 14, 25, 78, 108, 117–20, 122–24, 127–28, 130, 134–35
*Kolonien. See* colonies
Koolhaas, Rem, 142, 164
Krupp, Alfred, 25–26, 33, 44, 47, 53, 56–57, 59–61, 65–67, 70–73, 85, 92–94, 111, 159–60, 196, 214
Krupp, Friedrich, 33–34, 36, 49, 59–60
Krupp, Friedrich Alfred, 47
Krupp, Margarethe, 47, 59–61
Krupp Association for the Promotion of Mental and Social Culture (Krupp'sche Bildungsverein), 70

Landschaftspark Duisburg-Nord, 158, 161–62, 173, 176–77, 179, 182, 214
League of Small Animal Breeders, 129
*Lebensreform* movement. *See* nationalism

Le Corbusier, 84
Leeds (England), 35, 43
Lefebvre, Henri, 7, 10, 15, 112, 116
leisure, 94, 102–3, 133, 150; bourgeois, 129; democratic politics, as space for, 134; greening, 68; leisure education, 126–27; leisure landscape, 161–62; leisure society, 83, 85; leisure value, 121; as nature, 62, 66, 70, 73, 130, 206; parks, as oases of, 93, 96, 98
leisure spaces, 79, 92–93; as political spaces, 133–34
leisure time, 79, 83, 85, 126–27; as democratic, 96, 100, 104, 113; open space, 91–92
liberalism, 70; greening, compatibility with, 104
London (England), 10, 34–35, 44, 46–48, 203; Garden Bridge, 214, 216
Lowell (Massachusetts), 34, 43
Lüdtke, Hartmut, 79–80, 92–93, 96

management, 47, 147, 211; of colonies, 132; conflict, 127; democratic, 94; feudal style of, 81–82; impression, 169; labor, 24, 33; project, 164, 173; regional, 175, 194; soft, 163; Taylorist, 83; water, 49, 144, 165, 194
managerialism, 24, 171
Manchester (England), 35
Margarethenhöhe, 25, 47, 53, 65–66, 77, 83, 87, 94, 129, 132–33, 136, 153, 159–61, 214; animals, 56–57; cosmopolitanism of, 62–63; as garden city, 44, 54–55, 61; as gift, 59–61, 71; greenbelt, as transition zone, 60; greening, as social practice, 73–74; green space, 61; house rules, 56, 64, 147; as indirect good, 56; mind, cultivating of, 62–63; nature, different uses of, 57; nature, as space of leisure, 56, 62; new paternalism of, 69–70; noncompany residents, 61, 63; as philanthropic act, 72; as public good, 71; as public relations project, 59; rejuvenation, as area of, 62; streetcar system, 60; trees in, 57; urban infrastructure, 70;

urbanized nature, 56–57; as urban public space, 68
Marx, Karl, 7
Marxism, 7, 20, 112
mass culture, 84, 105, 115
mass media, 118
materialism, and culture, 201
materiality: of culture, 202; social imaginaries, 202
McKinsey and Company, 180–82
megacities, 2
Metzendorf, Georg, 54–55, 60–62, 93–94
Middle East, 1–2, 4, 17, 85, 203, 205–6
Mitscherlich, Alexander, 84, 115
modern art, 46
modernism: in architecture, 88, 115; urban, 115
modernity, 55, 154; modernist planning, 108
Moses, Robert, 88–89, 202
Moscow (Russia), 10

nationalism, 18, 73, 207; as modular, 19; nature, relationship between, 11; as social form, 19
national parks, 29
National Parks Service, 215
National Socialism, 11, 55, 78, 94–95, 98, 111, 121, 130, 190; urban public life, death of, 77. *See also* Third Reich
Native Americans, 212
nature, 47, 98, 104, 108–9, 118, 147, 153, 171, 203, 215–18; as aspirational, 56, 72–73, 78; as benevolent gift, 208; characteristics of, 23–24; and cities, 6–8, 201, 204; as civic space, 71; colonies, 39, 110, 113, 131; control, as mechanism of, 202; critique of, 195–96; cultural and materialist approaches to, 7; culture, mirror for, 7–8; and erasure, 211–12; as good, 25, 52, 200; good intentions, 212; green landscapes, 143; imaginaries of, 52; as indirect good, 56, 65, 72–73, 78; as leisure, 62, 66, 70, 73, 130, 206; managerial utility of, 73;

nature (cont.)
moral and spiritual improvement, 70–71; moral authority of, 21, 31, 65, 199; as moral benefit, 73; as moralized, 29; nationalism, relationship between, 11; neighborliness, connection between, 124–25; obscure, tendency to, 212; and politics, 134; public debate, 195, 199; as public good, 65–66, 82, 174–75, 179, 189–90, 194–96; public green space, 62; as public sphere, 65; romantic view of, 29–30, 59; signifiers of, 21, 56, 210; social, lying outside of, 23; social and political power, spatializations of, 11; social construction of, 202; social goods, 204; as social products, 72–73; social reproduction, 70; social value of, 217; as space of leisure, 56–57; subsistence-oriented form of, 41, 52, 56–57; symbolic value of, 31, 73; turn to, as imaginative shift, 31; universal benefits of, 24, 70–71, 73, 191, 199, 211; as universal good, 56, 72, 82, 104, 192–93, 213; universality of, 25, 78, 143; as unmediated experience, 166–67, 170; as urban, view of, 57; and urbanism, 13; and urbanization, 18, 54, 57, 129–30, 201, 206; urbanization, as constitutive essence, 17; urban problems, relief from, 29; urban publicness, 143; urban public space, improvement of, 29–30; as well-intentioned, 200; as word, 3, 15, 210. *See also* urbanized nature

Nazis. *See* National Socialism

Negt, Oskar, 14, 25, 78, 108, 117–20, 122–24, 127–28, 134–35

neoliberalism, 13, 23, 152; greening projects, 199; public sphere, dismantling of, 151

neoliberalization, 148, 151

New England, 29

New Left, 13, 25, 78, 108, 113, 116, 128, 153

New York City, 6, 10, 12, 17, 34–35, 48, 88, 202–3; Botanic Garden, 1; Brooklyn Bridge Park, 151, 192; Central Park, 4, 29, 192, 209–10, 214; Harlem, 209–10; High Line, 141, 148, 188, 192, 204, 213–14; Hudson Yards, 212–13; Manhattan, 160; Pier 55, 214; pocket parks, 216; poor doors, 213; South Bronx, 115; Washington Square Park, 166–67

North America, 2, 4, 17, 20, 54, 203, 205; postwar boom, 83

North Rhine–Westphalia (NRW), 9, 83, 87–88, 90, 102–3, 105, 129, 143, 145, 152

Oberhausen (Germany), 32, 94, 107, 120; Gasometer, 165, 173

Occupy Wall Street, 22

Olmsted, Frederick Law, 29, 210; urban elite, 209

Osthaus, Karl Ernst, 44, 46–49, 54, 56, 59, 61, 66, 71–72, 80, 85, 93–94, 111, 160, 214; garden city, 50–51

Paris (France), 10, 29, 34–35, 44, 46, 73, 148, 173–74, 202; Eiffel Tower, 159, 165; Parc André Citroën, 141

*Park and the People, The* (Rosenzweig and Blackmar), 209

parks, 206, 210; affordable housing, 216; as aspirational, 214; democracy, access to, 92, 96, 105, 209; discrimination, 215; as environmental hegemonies, 215; as gifts, 214; and leisure, 93, 96, 98; as public goods, 214; public sphere, 98; as revenue generators, 152; segregation, 215; as uncontroversial, 214; universal benefits, 209; and working class, 94–96. *See also* urban parks

Parson, Talcott, 84

paternalism, 104; company towns, 41; Eisenheim movement, 134, 136; feudal, form of, 41, 69; and greening, 24, 74, 212; and housing, 37; of labor management, 33, 37, 41; new paternalism, 69–70; urbanized nature, 196

phenomenology: of mining, 43; of nature experience, 98

Phoenix Lake, 162, 167, 172–75, 177–78, 182–85, 187–88, 191, 212, 214; criticism of,

195–96; development costs, 190; as fragile ecosystem, 192; and gentrification, 198; as greening project, 189, 193, 199–200; IBA, as break from, 180; rising rents, 194; shared imaginary, as signifier of, 193; as social project, 189; as structural change, 186; Turkish immigrants, 194, 196–98, 200; urban vision, 180; as "win-win," 213. *See also* Dortmund (Germany)

Phoenix Lake Development Corporation, 180

Pittsburgh (Pennsylvania), 10, 148

Poland, 33, 39, 98

polycentrism, 10, 36, 49, 68, 72, 154

preservation movement, 109–10

private space, 3, 25

Progressive movement, 29

Projekt Ruhr GmbH, 164

proletarian politics, 121; company housing, as haven for, 113; New Left, 116; urbanized nature, 129

public culture, 70; end of, 81

public education, 70

public good, 204–5; and greening, 69, 71, 208; greening projects, 23, 171, 187–88, 193–94, 199–200; as imperfect, 208–9; managerialism, acts of, 171; and nature, 65–66, 82, 174–75, 179, 189–90, 194–96; parks, 214; and trees, 66, 216; urbanized nature, 200, 208, 210–11

publicness, 80, 118; contested meanings of, 209

public politics, 79–80

public realm, 94; green space, 91–92

publics: diversification of, 104; and place, 80; public places, 80

public sphere, 3–4, 41, 70–71, 92, 104–5, 128, 131, 209–10; abstract social space, 80; bourgeois, 78, 80–81, 94, 100, 108, 117–20, 124, 135–36, 208; and democracy, 79–80, 83, 88–89, 98, 106, 153; democratic common good, 106; green space, 82; as half, 119, 122–23; inherited culture, 81; and nature, 65; neoliberalism, 151; proletarian, 23, 25, 117–20; spatial planning, 84–85; urban parks, 98; urban public space, 80–81. *See also* counterpublic sphere; urban public space

*Public Sphere and Experience* (Negt and Kluge), 117–19; spatial ideal, alternative version of, 120

public urban vision, imagination, critical role of, 52

Randstad (Holland), 10

Recklinghausen (Germany), 32

Regionalverband Ruhr (RVR), 14, 144, 156, 163

*Revierparks*, 25, 78, 82, 102, 107–8, 111, 113–14, 120, 123, 126, 131, 145, 147, 151–52, 159, 185–86, 190, 214; as bourgeois public sphere, 117, 138; brochure, 93–94, 96, 98–99, 103–5, 110, 112, 121, 124, 129, 160, 198; citizen participation, spatial analogue to, 92; deindustrialization, 134; and democracy, 92, 95, 103–4, 116; democracy, space-times of, 79–80; democratic public life, sites of, 79, 93; green space, 92; leisure, and democratic politics, 134; leisure oasis, demarcation of, 133; public sphere and democratic common good, representing of, 106; public sphere of, 128; structural change, 99–101; Turkish immigrants, 96, 198

Rhine-Ruhr (Germany), 10

Riis, Jacob, 35

Rocky Flats National Wildlife Refuge, 212

romanticism, in Germany, 11

Rosenzweig, Roy, 209

"Ruhrbanität," 10–11

Ruhrkohle AG (RAG), 132

Ruhr Museum, 157, 164, 176

Ruhr Valley, 11, 14, 17, 22, 73, 89, 90, 93, 108, 112, 126, 144–45, 155–56, 160, 163, 172–73, 175, 177–79, 182, 185, 196, 198; agrarian imaginaries, 52; air pollution, 101; animals in, 48, 101, 128, 130; arms production, role

Ruhr Valley (*cont.*)
in, 83; automobile oriented, 91; bourgeois, efforts to make, 111; changing demographics, 128; changing landscape, 150; city and hinterland, lack of division between, 36, 45; coal crisis, 85; coal mining in, 10, 32–34, 58, 77, 83, 98, 132, 149; codetermination policy, 132; company housing, 37–40, 95, 111; cosmopolitanism, lack of, 44, 46; deindustrialization of, 86–87, 100–101, 186; development program, 88; economic importance of, 43; factory closures in, 131–33, 141, 149; feudal management style, of barons, 81–82; foreign-born workforce, 72; French occupation of, 101, 130; garden cities of, 47–54, 59, 72, 82; "go-ins," 115; as green, 187; green, as good, 57; greening of, 65–66, 70, 72, 74, 77–78, 101, 141, 193, 202–7; greening projects in, 4–5, 9, 12–13, 20–21, 23, 30–31, 51, 148, 151; green space, 31, 50, 78, 128, 130; green space in industrial workers' housing, 24, 111; green space policy, 68–69; green urban public space, 104; higher education institutions, 42; high-rise apartment block, 91, 94; housing in, 30, 48, 81–82; housing crisis, 31, 34–37; immigrant population of, 96–98; as in-between-city, concept of, 10; industrial barons, waning of, 87; industrial economy, collapse of, 85, 87; industrial heritage of, 136–37; industrialization of, 36, 46–47, 180; industrial past, reinvention of, 142; intellectual elite, 42, 121, 129–30; landscape, as unmediated, 167; landscape beautification, 102–3; as leisure landscape, 161–62; middle class, 64; miners' strikes, 42–43; mining in, 10; morphology of, 36, 49, 51, 85, 193; nature, as antithesis of, 157; nature, as direct good, 25, 52; noncity urbanism, example of, 9; as paradoxical, 9; paternalism in, 24, 33; planning principles, 68–69; as polluted, 10; polycentrism of, 10, 36, 49, 72, 154; preservation, 137; recognized images, monumentalizing of, 165; re-envisioning of, 146–47, 161; river system, renaturalization of, 165–66; structural change, 88, 101, 132, 164–65; tourism- and consumption-based economy of, 137; transformation, of landscape, 153; universities in, 91, 94, 100; unsettled time in, 113; urban culture, lack of, 47; urban identity, 159; urbanism of, 30, 45–47, 49, 59, 159, 203; urbanized nature, 24–25, 37, 51–52, 54, 57, 59, 65; urbanizing of, 31–32, 38; urban parks, and working class, 94–96; water management, 49; working class in, 9–10, 44, 55, 57, 60, 63, 66, 87, 94, 98, 100, 114–16, 119, 128–31, 136, 165; working class, embourgeoisement of, 64. *See also* colonies; company housing; Eisenheim movement; *Revierparks*
rule of freedom, 70
rural studies, 7
Rust Belt, 4

Salin, Edgar, 84, 105
San Francisco (California), 160
Schmidt, Robert, 10, 36, 44–47, 51, 56, 61, 66, 71–72, 79, 85, 111, 144, 154–56, 163; greenbelts, 69, 90, 92, 146; green space policy, 68–69; urbanism, vision of, 49–50, 69
*Schrebergärten*, 128. *See also* allotment gardens
Schwarze-Rodrian, Michael, 144–47, 160, 162, 164, 169–70, 173, 178, 189; greenbelts, 156; soft management, 163
Second German Leisure Congress, 103
Sennett, Richard, 63, 81
Shanghai (China), 3
Siebel, Walter, 10, 145, 150, 152, 159
Siedlungsverband Ruhrkohlenbezirk (SVR), 45, 69, 79, 87–88, 90, 93–95, 102, 144; air pollution, concern over, 101; leisure time, and open space, 91–92
Sierau, Ulrich, 172, 185–86, 191
Sieverts, Thomas, 14, 152, 154, 169; *Zwischenstadt* (in-between-city) concept, 10, 155–56, 159

Silesia, 33
Sinclair, Upton, 58
Sitte, Camillo, 55
'68ers, 78, 113–16, 118, 134
smart cities, 1–2, 4, 203
social control, 40
Social Democratic Party (SPD), 42–43, 86, 114; disappointment in, 131–33
social imaginaries, 5, 18–19, 201, 206–7, 213, 215–16; and greening, 20, 65; green space, 78; materiality of, 202; modularity of, 73; social practices, 112; social projects, 174; urbanized nature, 22, 108, 211
social justice movements, 217
social mobility, high-rise apartment block, 113–14
social reform movements, 35, 57, 66, 114
sociology, 7, 21; historical, 12; social imaginary, 18; urban planning, relationship between, 85; and urban studies, 202
Sperber, Manès, 105
St. Louis (Missouri), Pruitt-Igoe complex, 115
*Structural Transformation of the Public Sphere, The* (Habermas), 80, 104, 154
subsistence agriculture, 51, 57
suburbia, criticism of, 115
sustainable agriculture, 217
sustainable cities, 1–2

Taylorism, 83
Third Reich, 206. *See also* National Socialism
Thompson, E. P., 10, 116
Thyssenkrupp, 33, 148, 180–81, 184. *See also* Friedrich Krupp AG
Tokyo (Japan), 10
*To-Morrow* (Howard), 47
tourism, 150
trees, 3–4, 15, 46, 58, 110, 121, 196, 210, 214; in American cities, 216; benefits of, 211; downside of, 216; in Margarethenhöhe, 57; planting of, 67, 99, 101–2, 166, 204; as public good, 66, 216; in urban streets, 19, 65, 67–68. *See also* green space

"Trouble with Wilderness, The" (Cronon), 211
Turkey, 98
Turkish immigrants, 96–97; in Eisenheim colonies, 128; Phoenix Lake, 194, 196–98, 200

UNESCO, 164
United Arab Emirates, 203
United States, 10, 14, 29, 31, 35, 51, 91, 106, 115, 160, 195, 209; company towns, 41; Jeffersonian vision, 34; new paternalism, 69; postwar planning in, 88–89; social movements, 114; trees in, 216
urban: constitutive essence, 17; nominal essence, 17; as word, 15
urban farms, 217
urban forestry, ecosystem of disservices, 216
urban greening, 1, 4, 26, 148, 151, 201; as contemporary, 3; as defined, 3; durability of, 3; ecological sustainability, 2; social logics of, 136; as social practice, 5, 8–9, 14–15; ubiquity of, 3. *See also* greening
urban green space, 6, 203; green-as-good logic, 2, 12, 216; Left, association with, 112. *See also* green space
Urban Improvement Act, 84
urbanism, 34, 152, 154, 206; and citizenship, 23; crises of, 11; and democracy, 116; discrete settlements, 10; garden city, 49–50; global, 207; green, 3, 26; high-rise model of, 114; and housing, 34; ideal, 1–2; industrial, 17; international ideals of, 23; middle class, 85; modern, 80–81; and nature, 13; noncity, 9; notorious moments of, 4; problems with, 5, 17, 21; public space, 80–81; rewilding efforts, 205; in Ruhr Valley, 30, 45, 49, 59, 159, 203; sustainable, 204, 217; walkable, 217. *See also* green cities
urbanity, 88, 103; decline of, 84; as defined, 120; mass culture, 84; open society, 95–96; public and private spheres, tension between, 89; as term, 10; urban parks, 105

urbanization, 6, 9, 19, 29, 34, 46–47, 51, 155, 205, 207; as category of analysis, 15; as constitutive essence, 17; epistemologies and imagination, 16; and greening, 206; greening practices, 17; and industrialization, 20, 30, 206; and nature, 18, 54, 57, 129–30, 201, 206; nature, as constitutive essence, 17; phenomena and condition, 16; product of, 52; of social consciousness, 16–17

urbanized nature, 5, 19, 26, 29–30, 37, 43, 54, 65, 71, 77, 109, 153, 179, 202, 205–6, 213, 218; in action, 25; animals, 15, 57; as aspirational, 21, 23, 25, 52–53, 59; as bourgeois, 129; in cities, 20; and citizenship, 73; as creative, 23; garden cities, 31, 51–53, 72; and greening, 71, 111–12, 207–8; greening practices, 15, 24; greening projects, 143; as imaginary of form, 22, 25; imaginative availability, 113; as indirect good, 25; managerial utility of, 73; nature, ideas about, 15; paternalism, 196; periodization of, 13–14; as proletarian, 129; public bathrooms, 217; as public good, 200, 208, 210–11; signifiers of, 56; as social imaginaries, 22, 108, 211; social organization of experience, erasure of, 170; universality of, 21, 25, 64; urban greening, 136; v. urban nature, 15; as well-intentioned action, 174. *See also* nature

urban modernism, 115

urban parks, 82; democratic mandate of, 92, 96–98, 105; folky names, 94–95; leisure, as oases of, 93, 96, 98; middle-class experiences of, 96; structural change, 100–101; town square, traditional functions of, 92; universal accessibility of, 104; urbanity, as model of, 105. *See also* parks; *Revierparks*

urban planning, 84, 92, 94, 103–4, 106, 141, 154, 205; functionalism in, 88–89; high-rise apartment block, 105; open space planning, 89; sociology, relationship between, 85

urban publicness, 151

urban public space, 60, 68, 77; fear of, 81; functional division of, 90–91; green space, 81; public sphere, 80–81; as universally accessible, 106; and urbanism, 80–81. *See also* public sphere

urban renewal, 88

urban studies, 7, 15–16, 155, 202

urban sustainability, 141; green urbanism, 3

van Gogh, Vincent, 46
Vaux, Calvert, 29
Venice (Italy), 160
Vidal, Fernando, 21
Vienna (Austria), 173–74
*Volk* (people), 79
*Volksparks* (people's parks), 79

Weber, Max, 10
Werkbund, 44
West Berlin (Germany), 143
West Germany, 9, 11, 77, 79–81, 85, 87–90, 95–96, 103, 105, 114, 118; automobile-oriented, 91; greening projects in, 78; housing crisis, 84; leisure, phenomenon of, 83. *See also* East Germany; Germany
Westphalian Industrial Museum, 137
wilderness, 211–12; discrimination, 215; segregation, 215
Wilhelm II, 42, 61
Williams, Raymond, 3, 7, 10, 12–13, 112, 116
Wirth, Louis, 18
Wolfsburg (Germany), 181
working class, 30, 40–42, 46, 65, 79, 91, 197, 208; class consciousness, 116, 131; culture, 113, 131, 165; depoliticization, worry about, 120; disappearance of, 136; in Eisenheim colony, 113, 120, 131, 133–36; green spaces, 130; identity, 9; and parks, 94–96; in Ruhr Valley, 9–10, 44, 55, 57, 60, 63–64, 66, 87, 94, 98, 100, 114–16, 119, 128–31, 136, 165
*World Cities, The* (Hall), 10
World's Columbian Exposition, 44
World War I, 42, 130

World War II, 33, 77, 79, 82–83, 94, 98, 101, 103, 132
*Wulfen Coal Mine* (Hebig), 167

Yugoslavia, 98

Zeche Zollern colliery, 137, 164–66
Zille, Heinrich, 35
Zöpel, Christoph, 156
*Zwischenstadt* (in-between-city) concept, 10, 155–56, 159